International He
Safety at Work I

This companion to the bestselling *International Health and Safety at Work* will help you prepare for the written assessments on the NEBOSH International General Certificate in Occupational Health and Safety (March 2011 specification).

It provides complete coverage of the syllabus in bite-sized chunks and will help you learn and memorise the most important areas, with links provided back to the main *International Health and Safety at Work* text to help you consolidate your learning.

● Small and portable making it ideal for use anywhere: at home, in the class-room or on the move

● Includes specimen questions and answers from recent examination papers

● Everything you need for productive revision in one handy reference source

This revision guide is written by an experienced lecturer who has spent many years helping students become accredited by NEBOSH. Ed Ferrett is former Vice Chairman of NEBOSH (1999–2008) and a lecturer on NEBOSH courses with both public and private course providers. He is a Chartered Engineer and Health and Safety Consultant.

International Health and Safety at Work Revision Guide

For the NEBOSH International General Certificate

Ed Ferrett

Routledge
Taylor & Francis Group

LONDON AND NEW YORK

Published 2012 by Routledge
2 Park Square, Milton Park, Abingdon, Oxon OX14 4RN

Simultaneously published in the USA and Canada
by Routledge
711 Third Ave, New York, NY 10017

Routledge is an imprint of the Taylor & Francis Group, an informa business

British Library Cataloguing in Publication Data
A catalogue record for this book is available from the British Library

Library of Congress Cataloging in Publication Data
   Ferrett, Ed.
   International health and safety at work revision guide / by Ed Ferrett.
      p. ; cm.
   Companion to: Introduction to international health and safety at work / Phil Hughes, Ed
   Ferrett. 2010.
   I. Hughes, Phil, MSc, FIOSH, RSP. Introduction to international health and safety at work.
   II. Title.
   [DNLM: 1. Occupational Health–Outlines.   2. Accidents, Occupational–prevention &
   control–Outlines.   3. Safety Management–Outlines. WA 18.2]
   363.11–dc23

                                                                                    2011050659

Pbk ISBN13: 978-0-415-51980-9
Ebk ISBN13: 978-0-203-11374-5

Typeset in Rotis Sans Serif
by RefineCatch Ltd, Bungay, Suffolk

Contents

Preface

Welcome to the International *Health and Safety at Work Revision Guide: for the NEBOSH International General Certificate*. The Guide has been designed to be used together with the NEBOSH International General Certificate syllabus and the textbook *International Health and Safety at Work* by Hughes and Ferrett. The guide gives only a basic summary of the NEBOSH International General Certificate course and a full explanation of all the topics is given in the textbook.

The Revision Guide has the following features:

- It follows the latest NEBOSH International General Certificate syllabus;

- Revision notes for each of the elements of the two units – IGC1 (Management of International Health and Safety) and IGC2 (Controlling International Workplace Risks);

- A summary of the learning outcomes and key points is given for each element;

- Important diagrams are included to help revision;

- There is a final section to advise on preparation for the examination and some specimen answers to both long and short answer questions taken from recent NEBOSH examination papers.

The Revision Guide will also be useful to those who work outside the UK and have specific health and safety responsibilities in their jobs and those who are studying on other courses that include important health and safety elements, for example courses in Engineering, Business Studies, the Health Services and Retail and Hotel Management.

The compact size of the Revision Guide ensures that it can be easily carried and used for revision at any time or place. It should be used throughout the course alongside the textbook and course hand-outs.

Good luck with your studies.

List of Principal Abbreviations for International Guide

ACGIH	American Conference of Governmental Industrial Hygienists
ACM	Asbestos containing material
AFFF	Aqueous film forming foam
COSHH	Control of Substances Hazardous to Health
dB	Decibels
dB(A)	Decibel (A-weighted)
dB(C)	Decibel (C-weighted)
DSE	Display screen equipment
EU	European Union
HAV	Hand–arm vibration
HSE	UK Health and Safety Executive
ILO	International Labour Organization
LEV	Local exhaust ventilation
LPG	Liquefied petroleum gas
LTEL	Long-term exposure limit
MEL	Maximum exposure limit
MEWP	Mobile elevated work platform
MIRA	Motor Industry Research Association
MSD	Musculoskeletal disorder
NEBOSH	National Examination Board in Occupational Safety and Health
OSH	Occupational Safety and Health
OES	Occupational exposure standard
PPE	Personal protective equipment
RCD	Residual current device
RPE	Respiratory protective equipment
SPL	Sound pressure level
STEL	Short-term exposure limit

TLV	Threshold limit value
TWA	Time-weighted average
WBV	Whole body vibration
WEL	Workplace exposure limit
WRULD	Work-related upper limb disorder

Management of
International Health
and Safety

Foundations in Health and Safety

1.1

LEARNING OUTCOMES

Outline the scope and nature of occupational health and safety

Explain the moral, social and economic reasons for maintaining and promoting good standards of health and safety in the workplace

Explain the role of national governments and international bodies in formulating a framework for the regulation of health and safety.

 KEY REVISION POINTS

The definitions of hazard and risk ☐

The business case for health and safety (direct, indirect, insured and uninsured costs) ☐

The health and safety responsibilities and duties of employers to their employees and others affected by their undertaking ☐

The health and safety rights and responsibilities of workers ☐

The role of international organisations in setting international standards for health and safety at work ☐

Sources of information on national standards ☐

DEFINITIONS

- ▪ Accident – an unplanned event that results in damage, loss or harm

- ▪ Hazard – the potential of something to cause harm

- ▪ Risk – the likelihood of something to cause harm

- ▪ Residual risk – remaining risks after controls applied

- ▪ Welfare – provision of facilities to maintain health and well-being of people in the workplace (e.g. washing, sanitary facilities and first aid)

- ▪ Near miss – any incident that could have resulted in an accident

- ▪ Dangerous occurrence – a near miss that could have led to serious injury or loss of life.

REASONS FOR GOOD HEALTH AND SAFETY MANAGEMENT

MORAL REASONS

Need to provide a reasonable standard of care and ethical reasons to reduce:

■ accident rates

■ industrial disease and ill-health rates.

SOCIAL REASONS

These include:

■ societal expectation of good standards of health and safety

■ a duty of care – to provide:

- a safe place of work, including access and egress

- safe plant and equipment

- a safe system of work

- safe and competent fellow employees; and

- adequate levels of supervision, information, instruction and training.

ECONOMIC REASONS

Poor health and safety management can lead to:

■ direct and

■ indirect costs.

Good health and safety management can lead to:

■ a more highly motivated workforce, resulting in an improvement in the rate of production and product quality;

■ improved image and reputation of the organisation with its various stakeholders.

COSTS OF ACCIDENTS AND ILL HEALTH

DIRECT COSTS

Directly related to the accident and may be insured or uninsured.

Insured direct costs normally include:

- claims on employers' and public liability insurance;

- damage to buildings, equipment or vehicles; and

- any attributable production and/or general business loss.

Uninsured direct costs include:

- fines resulting from prosecution by the enforcement authority;

- sick pay;

- some damage to product, equipment, vehicles or process not directly attributable to the accident (e.g. caused by replacement staff);

- increases in insurance premiums resulting from the accident;

- any compensation not covered by the insurance policy due to an excess agreed between the employer and the insurance company; and

- legal representation following any compensation claim.

INDIRECT COSTS

Costs which may not be directly attributable to the accident but may result from a series of accidents.

Insured indirect costs include:

- a cumulative business loss;

- product or process liability claims; and

- recruitment of replacement staff.

Uninsured indirect costs include:

- loss of goodwill and a poor corporate image;

- accident investigation time and any subsequent remedial action required;

- production delays;

- extra overtime payments;

- lost time for other employees, such as a First Aider, who attend to the needs of the injured person;

- the recruitment and training of replacement staff;

- additional administration time incurred;

- first aid provision and training; and

- lower employee morale, possibly leading to reduced productivity.

Some of these items, such as business loss, may be uninsurable or too prohibitively expensive to insure. Therefore, insurance policies can never cover all of the costs of an accident or disease, because either some items are not covered by the policy or the insurance excess is greater than the particular item cost.

EMPLOYERS' LIABILITY COMPULSORY INSURANCE

- covers the employer's liability in the event of accidents and work-related ill health to employees and others who may be affected by their operations;

- ensures that any employee who successfully sues his/her employer following an accident is assured of receiving compensation irrespective of the financial position of the employer; and

- is a legal duty in many countries.

THE ROLE OF NATIONAL GOVERNMENTS AND INTERNATIONAL BODIES IN FORMULATING A FRAMEWORK FOR THE REGULATION OF HEALTH AND SAFETY

EMPLOYERS' RESPONSIBILITIES

The principal general duties of employers under the ILO Recommendation 164 are:

(a) to provide and maintain workplaces, machinery and equipment, and use work methods, which are as safe and without risk to health as is reasonably practicable;

(b) to give necessary instruction and training that takes into account the functions and capabilities of different categories of workers;

(c) to provide adequate supervision of work practices, ensuring that proper use is made of relevant occupational health and safety measures;

(d) to institute suitable occupational health and safety management arrangements appropriate to the working environment, the size of the undertaking and the nature of its activities; and

(e) to provide, without any cost to the worker, adequate personal protective clothing and equipment which are reasonably necessary when workplace hazards cannot be otherwise prevented or controlled.

WORKERS RESPONSIBILITIES AND RIGHTS

Workers' rights are embodied in the ILO 'Declaration on Fundamental Principles and Rights at Work':

1 Freedom of association – the right of workers and employers to form and join organisations of their choice.

2 Freedom from forced labour.

3 Freedom from discrimination.

4 Opposition to child labour.

The ILO Code of Practice – 'Ambient factors in the workplace' – specifies that workers and their representatives should have the right to:

(a) be consulted regarding any hazards or risks to health and safety from hazardous factors at the workplace;

(b) inquire into and receive information from the employer regarding any hazards or risks to health and safety from hazardous factors in the workplace;

(c) take adequate precautions, in co-operation with their employer, to protect themselves and other workers against hazards or risks to their health and safety;

(d) request and be involved in the assessment of hazards and risks to health and safety by the employer and/or the competent authority, and in any subsequent control measures and investigations;

(e) be involved in the inception and development of workers' health surveillance, and participate in its implementation; and

(f) be informed in a timely, objective and comprehensible manner:

  (i) of the reasons for any examinations and investigations relating to the health hazards involved in their workplace;

  (ii) individually of the results of medical examinations, including pre-assignment medical examinations, and of the subsequent assessment of health.

In accordance with national laws and regulations, workers should have the right:

(a) to bring to the attention of their representatives, employer or competent authority any hazards or risks to health and safety at the workplace;

(b) to appeal to the competent authority if they consider that the measures taken and the means employed by the employer are inadequate for the purposes of ensuring health and safety at work;

(c) to remove themselves from a hazardous situation when they have good reason to believe that there is an imminent and serious risk to their health and safety and inform their supervisor immediately;

(d) in the case of a health condition, such as sensitisation, to be transferred to alternative work that does not expose them to that hazard, if such work is available and if the workers concerned have the qualifications or can reasonably be trained for such alternative work;

(e) to compensation if the case referred to in (d) above results in loss of employment;

(f) to adequate medical treatment and compensation for occupational injuries and diseases resulting from hazards at the workplace; and

(g) to refrain from using any equipment or process or substance which can reasonably be expected to be hazardous, if relevant information is not available to assess the hazards or risks to health and safety.

WORKERS' RESPONSIBILITIES

Employees or workers have specific responsibilities under the ILO Convention 187, which are to:

(a) take reasonable care for their own safety and that of other persons who may be affected by their acts or omissions at work;

(b) comply with instructions given for their own health and safety and those of others and with health and safety procedures;

(c) use safety devices and protective equipment correctly and not to render them inoperative;

(d) report forthwith to their immediate supervisor any situation which they have reason to believe could present a hazard and which they cannot themselves correct; and

(e) report any accident or injury to health which arises in the course of or in connection with work.

THE ROLE OF ENFORCEMENT AGENCIES

The national legal framework usually consists of:

■ Acts or other statutory instruments;

■ Regulations; and

■ Codes of Practice, Guidance and Standards.

REGULATORY AUTHORITIES AND ENFORCEMENT AGENCIES

Many national governments relate the level of regulatory inspections to the effectiveness of the health and safety management system adopted by the organisation. Several countries have adopted the ILO-OSH 2001 system.

It is likely that more countries and multinational companies will expect occupational health and safety management systems to be adopted either as a legal duty or an implied duty following rulings from local courts of law.

CONSEQUENCES OF NON-COMPLIANCE

- Loss of competitive advantage

- Inability to compete for certain contracts

- Fines

- Imprisonment.

INTERNATIONAL STANDARDS AND CONVENTIONS

RELEVANT INTERNATIONAL ORGANISATIONS

1 ISO (International Organization for Standardization) has published 18 500 International Standards. Those of importance in health and safety are:

- ISO 9000:2005 – Quality management systems

- ISO 14001:2004 – Environmental management systems.

2 The International Labour Organization (ILO) has many Conventions, Recommendations and Codes of Practice. Its occupational safety and health standards can be divided into four groups, and an example is given in each case:

1 *Guiding policies for action*

The Occupational Safety and Health Convention, 1985 (No. 155) and its accompanying Recommendation (No. 164) emphasise the need for preventative measures and a coherent national policy on occupational safety and health. They also stress employers' responsibilities and the rights and duties of workers.

2 *Protection in given branches of economic activity*

The Safety and Health in Construction Convention, 1988 (No. 167) and its accompanying Recommendation (No. 175) stipulate the basic principles and measures to promote safety and health of workers in construction.

3   *Protection against specific risks*

The Asbestos Convention, 1986 (No. 162) and its accompanying Recommendation (No. 172) give managerial, technical and medical measures to protect workers against asbestos dust.

4   *Measures of protection*

Migrant Workers (Supplementary Provisions) Convention, 1975 (No. 143) aims to protect the safety and health of migrant workers.

The ILO has a health and safety management system (ILO-OSH 2001), shown in Figure 1 in the next element (Policy).

SOURCES OF INFORMATION ON NATIONAL STANDARDS

■ Health and safety legislation;

■ HSE publications, such as Approved Codes of Practice, guidance documents, leaflets, journals, books and their website;

■ European and British Standards;

■ International Labour Organization (ILO);

■ Occupational Safety and Health Administration (USA);

■ European Agency for Safety and Health (EU);

■ Worksafe (Western Australia);

■ Health and safety magazines and journals;

■ Information published by trade associations, employer organisations and trade unions;

■ Specialist technical and legal publications;

■ Information and data from manufacturers and suppliers; and

■ The internet and encyclopaedias.

Health and Safety Management Systems – Policy

1.2

LEARNING OUTCOMES

Outline the key elements of a health and safety management system

Explain the purpose and importance of setting policy for health and safety

Describe the key features and appropriate content of an effective health and safety policy

KEY REVISION POINTS

The seven key elements of a health and safety management system ☐

The three key elements of a health and safety policy ☐

Target setting for health and safety performance ☐

Circumstances leading to the need for a review of health
and safety policy ☐

The reasons for unsuccessful health and safety policies ☐

The standards and guidance relating to health and safety policies ☐

KEY ELEMENTS OF A HEALTH
AND SAFETY MANAGEMENT SYSTEM

The seven key elements are:

1 **Policy** – A clear health and safety policy contributes to business efficiency and continuous improvement throughout the operation. The policy should state the intentions of the organisation in terms of clear aims, objectives, targets and senior management involvement.

2 **Organising** – A well-defined health and safety organisation should identify health and safety responsibilities at all levels of the organisation. An effective organisation will be noted for good communication, the promotion of competency, the commitment of all employees and a responsive reporting system.

3 **Planning and implementing** – A clear health and safety plan, based on risk assessment, sets and implements performance standards, targets and procedures through an effective health and safety management system. The plan should set priorities and objectives for the control or elimination of hazards and the reduction of risks.

4 **Measuring performance** – This includes both active (sometimes called proactive) and reactive monitoring of the health and safety management system. It is also important to measure the organisation against its own long-term goals and objectives.

5 **Reviewing performance** – The results of monitoring and independent audits should indicate whether the objectives and targets set in the health and safety policy need to be changed. Changes in the health and safety environment in the organisation, such as an accident, should also trigger a performance review. Performance reviews should include comparisons with internal performance indicators and the external performance indicators of similar organisations with exemplary practices and high standards.

6 **Auditing** – An independent and structured audit of all parts of the health and safety management system reinforces the review process. If the audit is to be really effective, it must assess both the compliance with stated procedures and the performance in the workplace. It will identify weaknesses in the health and safety policy and procedures and identify unrealistic or inadequate standards and targets

7 **Continual improvement** – The simplest way to achieve continual improvement is to implement the recommendations of audits and management reviews and use benchmarks from similar organisations and any revised national or industrial guidelines. Suggestions from the workforce, managers, supervisors and the health and safety committee can be a very effective vehicle for continual improvement.

Figure 1 The framework for Health and Safety Management ILO-OSH 2001

PURPOSE AND IMPORTANCE OF SETTING POLICY FOR HEALTH AND SAFETY

The purpose of setting a health and safety policy is to ensure that:

- everybody associated with the organisation is aware of its health and safety aims and objectives and how they are to be achieved;

- the performance of the organisation is enhanced in areas other than health and safety;

- there is effective personal development of the workforce;

- business efficiency is improved throughout the operation; and

- the involvement of senior management in health and safety issues is evident to all stakeholders.

The policy should state the intentions of the business in terms of clear aims, objectives, organisation, arrangements and targets for all health and safety issues.

The written health and safety policy should include the following three sections:

- a health and safety **policy statement of intent** which includes the health and safety aims and objectives of the organisation;

- the health and safety **organisation** detailing the people with specific health and safety responsibilities and their duties; and
- the health and safety **arrangements** in place in terms of systems and procedures.

Policy statement of intent:

- Aims and objectives;

- Duties of employer and employees;

- Performance targets and benchmarks;

- Name of person responsible for health and safety; and

- Posted and dated.

Organisation:

- Safety manual
- Organisational chart
- Responsibilities
- Allocation of resources, including finance
- Safety monitoring system
- Identification of main hazards.

Arrangements to include:

- Planning and organising
- Accident reporting
- Emergencies
- Contractors and visitors
- Consultation and communication with employees
- Fire precautions
- Main risk assessments and hazard control
- Performance monitoring.

REASONS FOR A REVIEW OF THE HEALTH AND SAFETY POLICY

- Significant organisational and/or technological changes have taken place.
- There have been changes in personnel and/or legislation.
- Health and safety performance has fallen below the occupational group's benchmarks.
- The monitoring of risk assessments and/or accident/incident investigations indicate that the health and safety policy is no longer totally effective.
- Enforcement action has been taken by the national health and safety enforcement agency.
- A sufficient period of time has elapsed since the previous review.

THE EFFECTS OF A POSITIVE HEALTH AND SAFETY PERFORMANCE

Are to:

■ support the overall development of personnel;

■ improve communication and consultation throughout the organisation;

■ minimise financial losses due to accidents and ill health and other incidents;

■ directly involve senior managers in all levels of the organisation;

■ improve supervision, particularly for young persons and those on occupational training courses;

■ improve production processes; and

■ improve the public image of the organisation or company.

THE REASONS FOR UNSUCCESSFUL HEALTH AND SAFETY POLICIES

Include:

■ the statements in the policy and the health and safety priorities are not understood by, or properly communicated to, the workforce;

■ minimal resources are made available for the implementation of the policy;

■ too much emphasis on rules for employees and too little on management policy;

■ a lack of parity with other activities of the organisation (such as finance);

■ lack of senior management involvement in health and safety;

■ inadequate personal protective equipment;

■ unsafe and poorly maintained machinery and equipment; and

■ a lack of health and safety monitoring procedures.

STANDARDS AND GUIDANCE RELATING TO HEALTH AND SAFETY POLICY

The ILO recommends (in ILO-OSH 2001) that a health and safety policy should be:

(a) specific to the organisation and appropriate to its size and the nature of its activities;

(b) concise, clearly written, dated and made effective by the signature or endorsement of the employer or the most senior accountable person in the organisation;

(c) communicated and readily accessible to all persons at their place of work;

(d) reviewed for continuing suitability; and

(e) made available to relevant external interested parties, as appropriate.

Further, the ILO recommends that the policy should include, as a minimum, the following key principles and objectives:

(a) the protection of the health and safety of all members of the organisation by the prevention of work-related injuries, ill health, diseases and incidents;

(b) the compliance with relevant national laws and regulations, voluntary programmes and collective agreements involving occupational health and safety;

(c) consultation with the workers and their representatives and encouragement for them to participate actively in all elements of the health and safety management system; and

(d) the continual improvement of the performance of the health and safety management system.

Health and Safety Management Systems – Organising

LEARNING OUTCOMES

Outline the health and safety roles and responsibilities of employers, managers, supervisors, workers and other relevant parties

Explain the concept of health and safety culture and its significance in the management of health and safety in an organisation

Outline the human factors which influence behaviour at work in a way that can affect health and safety

Explain how health and safety behaviour at work can be improved

Outline the need for emergency procedures and the arrangements for contacting emergency services

Outline the requirements for, and effective provision of, first aid in the workplace.

KEY REVISION POINTS

The health and safety roles and responsibilities of employers to their employees and others affected by their undertaking ☐

The health and safety roles and responsibilities of directors, managers and supervisors ☐

The health and safety roles and responsibilities of workers ☐

Duties and responsibilities of manufacturers and others in the supply chain ☐

The responsibilities of clients and their contractors ☐

The competence and responsibilities of the health and safety practitioner/adviser ☐

The definition and importance of a health and safety culture ☐

The relationship between culture and performance ☐

The definition of human factors and their influence on the culture ☐

The development of a positive health and safety culture ☐

The importance of good communication ☐

The different forms of health and safety training ☐

The internal and external influences on the health and safety culture of an organisation ☐

The development of emergency and first aid procedures in the workplace ☐

ORGANISATIONAL HEALTH AND SAFETY ROLES AND RESPONSIBILITIES OF EMPLOYERS, DIRECTORS, MANAGERS, WORKERS AND OTHER RELEVANT PARTIES

ORGANISATIONAL HEALTH AND SAFETY RESPONSIBILITIES – EMPLOYERS

The organisational health and safety responsibilities of employers are to safeguard the health and safety and welfare at work of:

- employees;
- other workers – agency, temporary or casual;
- trainees;
- contractors;
- visitors; and
- neighbours and the general public.

Key actions required of the employer

- ensure competent advice on health and safety matters;
- obtain Employers' Liability insurance;
- compile a health and safety policy and ensure that an adequate health and safety management system is in place;
- ensure that risk assessments of all work activities are undertaken and any required controls are put in place;
- provide the workforce with health and safety information and training;
- provide adequate welfare facilities;
- consult the workforce on health and safety issues; and
- report and investigate some accidents, diseases and dangerous occurrences.

Principal general duties of employers recommended by the ILO

- to provide and maintain workplaces, machinery and equipment;
- to use safe work methods without risk to health as is reasonably practicable;

- to give adequate instructions and training;

- to provide adequate supervision;

- to institute organisationally relevant health and safety arrangements;

- to provide, without any cost to the worker, adequate personal protective clothing and equipment;

- to keep abreast of relevant scientific and technical knowledge;

- to consult with workers or their representatives;

- to set out a relevant health and safety policy in writing (where the size and nature of the organisation makes this appropriate);

- to keep records (such as incidences of accidents and ill health at work) as required by local legislation, and report these as required to the authorities concerned;

- to monitor the implementation of applicable standards;

- to employ specialists to advise on particular occupational safety or health problems or supervise the application of measures to meet them;

- to develop a preventative safety and health culture; and

- to conduct and record risk assessments.

Other employer responsibilities

Visitors and the general public

Possible hazards	Possible controls
• unfamiliarity with the workplace processes • lack of knowledge of site layout • unfamiliarity with emergency procedures • inappropriate personal protective equipment • inadequate or unsigned walkways • added vulnerability if young or disabled visitors	• visitor identification (use of badges) • routine signing in and out • escorted by a member of staff • provision of information on hazards and emergency procedures • given explicit site rules (wearing of PPE) • clear marking of walkways

For night workers, employers should:

- determine the normal working time for night workers;

- determine who may work at night – some vulnerable people, such as those under the age of 18 years and pregnant women, are excluded;

- if the working time is more than 8 hours per day on average, determine whether the amount of hours can be reduced or if any exceptions apply;

- develop an appropriate health assessment and method of making health checks;

- ensure that proper records of night workers are maintained, including details of health assessments; and

- ensure that night workers are not involved in work which is particularly hazardous.

ORGANISATIONAL HEALTH AND SAFETY RESPONSIBILITIES – DIRECTORS

Effective health and safety performance comes from the top and directors have both collective and individual responsibility for health and safety.

Directors and Board members must ensure that	Management of health and safety at Board level involves
• the health and safety of employees and others, such as members of the public, is protected • risk management includes health and safety risks and becomes a key business risk in board decisions • health and safety duties imposed by legislation are followed	• planning the direction of health and safety • delivering the plan for health and safety • monitoring health and safety performance • reviewing health and safety performance

ORGANISATIONAL HEALTH AND SAFETY RESPONSIBILITIES – MANAGERS

1. Managing Directors / Chief Executives

Managing Directors / Chief Executives are responsible for:

- the health and safety performance within the organisation;

- ensuring that adequate resources are available for the health and safety requirements within the organisation;

- the establishment, implementation and maintenance of a health and safety programme for the organisation that encompasses all areas of significant health and safety risk;

- the approval, introduction and monitoring of all site health and safety policies, rules and procedures; and

- the review and possible revision annually of the effectiveness of the health and safety programme.

2. Departmental managers

The principal departmental managers may report to the Site Manager, Managing Director or Chief Executive. In particular, they:

- are responsible and accountable for the health and safety performance of their department;

- must ensure that any machinery, equipment or vehicles used within the department are maintained, correctly guarded and meet agreed health and safety standards. Copies of records of all maintenance, statutory and insurance inspections must be kept by the Departmental Manager;

- develop a training plan that includes specific job instructions for new or transferred employees and follow up on the training by supervisors. Copies of records of all training must be kept by the Departmental Manager; and

- personally investigate all lost workday cases and dangerous occurrences and report to their line manager. Progress any required corrective action.

3. Supervisors

The supervisors are responsible to and report to their Departmental Manager. In particular, they:

- are responsible and accountable for their team's health and safety performance;
- enforce all safe systems of work procedures that have been issued by the Departmental Manager;
- instruct employees in relevant health and safety rules, make records of this instruction and enforce all health and safety rules and procedures;
- enforce personal protective equipment requirements, check that it is being used and periodically appraise condition of equipment; and
- record any infringements of the personal protective equipment policy.

Health and safety adviser must:

- be competent following the attainment of a health and safety qualification and training;
- report directly to a senior management on matters of policy;
- keep up to date with technological advances and legislative changes;
- advise on the establishment of health and safety, maintenance and accident investigation procedures; and
- provide liaison with external agencies, such as the HSE, Fire Authorities, contractors, insurance companies and the public.

EMPLOYEE RESPONSIBILITIES

Employees or workers have specific responsibilities under the ILO Convention, which are to:

- take reasonable care for their own safety and that of other persons who may be affected by their acts or omissions at work;
- comply with instructions given for their own safety and health and those of others and with safety and health procedures;
- use safety devices and protective equipment correctly and do not render them inoperative;

■ report forthwith to their immediate supervisor any situation which they have reason to believe could present a hazard and which they cannot themselves correct;

■ report any accident or injury to health which arises in the course of or in connection with work.

The self-employed are responsible for:

■ their own health and safety;

■ ensuring that others who may be affected are not exposed to risks to their health and safety;

■ undertaking risk assessment;

■ co-operating with others who work in the premises and, where necessary, in the appointment of a health and safety co-ordinator;

■ providing relevant information to other employees working in the premises.

PERSONS IN CONTROL OF PREMISES

The duty of 'Persons in control of **non-domestic** premises' extends to:

■ people entering the premises to work;

■ people entering the premises to use machinery or equipment;

■ access to and exit from the premises; and

■ corridors, stairs, lifts and storage areas.

DUTIES AND RESPONSIBILITIES OF MANUFACTURERS AND OTHERS IN THE SUPPLY CHAIN

Persons who design, manufacture, import or supply any article or substance for use at work to ensure, so far as is reasonably practicable, that it is safe and without risk to health. In addition:

■ articles must be safe when they are set, cleaned, used and maintained;

■ substances must be without risk to health when they are used, handled, stored or transported;

■ testing and examination must be carried out to ensure the required level of safety; and

■ employers should be provided with information on the safe use, dismantling and disposal of the articles and substances and given revised information should a subsequent serious risk become known.

ILO-OSH 2001 requires that procurement procedures adopted should ensure that:

■ compliance with health and safety requirements for the organisation is identified, evaluated and incorporated into purchasing and leasing specifications;

■ national laws and regulations and the organisation's health and safety requirements are identified prior to the procurement of goods and services; and

■ arrangements are made to achieve conformance to the requirements prior to their use.

Importers have a duty to ensure articles or substances comply with the requirements of national legislation.

ADVANTAGES OF GOOD SUPPLY CHAIN MANAGEMENT

■ reduction of waste of materials and time;

■ ability to respond rapidly to changing requirements at short notice;

■ reduction in accidents due to close liaison;

■ legal duties addressed satisfactorily; and

■ easier to maintain good customer care.

RESPONSIBILITIES OF CLIENTS AND THEIR CONTRACTORS

The following points should be considered when contractors are employed:

■ Health and safety must be included in the contract specification.

■ All significant hazards must be included in the contract specification.

- The contractor must be selected with safety in mind.

- Prior to the start of work, health and safety policies should be exchanged.

- The contractor must be given basic site and health and safety information, such as welfare and first aid arrangements, significant hazards, safe storage of chemicals and the name of the contract supervisor.

- Where appropriate, the contractor should supply risk assessments and method statements.

- For construction work, the contractor should be aware of the position of buried/overhead services and the arrangements for the disposal of waste.

- The contractor should be monitored during the progress of the contract by the contract supervisor.

- The contract supervisor should check that the work has been completed safely at the end of the contract.

SAFETY RULES FOR CONTRACTORS

Contractors' safety rules should contain as a minimum the following points:

- **Health and safety** – that the contractor operates to at least the minimum legal standard and conforms to accepted industry good practice;

- **Supervision** – that the contractor provides a good standard of supervision of their own employees;

- **Sub-contractors** – that they may not use sub-contractors without prior written agreement from the organisation; and

- **Authorisation** – that each employee must carry an authorisation card issued by the organisation at all times while on site.

CONSTRUCTION PROJECTS

Good practice on construction projects requires the following details to be checked by clients:

- the competence of all their appointees;

- there are suitable management arrangements for the project;

- there is sufficient time and resources for all stages; and
- pre-construction information is available to designers and contractors.

Contractors must:

- plan, manage and monitor their own work and that of employees;
- check the competence of all their appointees and employees;
- train their own employees;
- provide information to their employees;
- comply with the requirements for health and safety on site detailed in local legislation; and
- ensure there are adequate welfare facilities for their employees.

Clients should ensure that contractors provide:

- information regarding their health and safety policy;
- information on the contractor's health and safety organisation detailing the responsibilities of individuals;
- information on the contractor's procedures and standards of safe working;
- relevant method statements; and
- procedures for accident investigation.

SELECTION OF CONTRACTORS

The following issues should be considered:

- an adequate health and safety policy;
- competent supervision;
- the availability of competent health and safety advice;
- past accident record;
- ability to assess hazards and risks involved in the contract and implement appropriate control measures;

- examples of method statements;

- a competent, trained and experienced workforce; and

- good references from previous contracts.

JOINT OCCUPATION OF PREMISES

Where premises are occupied jointly, each employer should:

- satisfy themselves that the arrangements adopted for health and safety are adequate;

- co-operate and co-ordinate any necessary control procedures with other employers;

- take reasonable steps to co-ordinate with other employers to comply with legal requirements;

- agree joint arrangements to meet regulatory obligations, such as appointing a health and safety co-ordinator; and

- take reasonable steps to inform other employers where there are risks to health and safety.

CONCEPT OF HEALTH AND SAFETY CULTURE

The safety culture of an organisation is the product of individual and group values, attitudes, perceptions, competencies and patterns of behaviour that determine the commitment to, and the style and proficiency of, an organisation's health and safety management.

FEATURES OF A GOOD HEALTH AND SAFETY CULTURE

- leadership and commitment to health and safety at all levels;

- acceptance that high standards are achievable;

- mutual trust throughout the organisation;

- detailed risk assessments and control and monitoring procedures;

- health and safety policy including a code of practice and required health and safety standards;

- training, communication and consultation systems;

- encouragement to the workforce to report potential hazards;

- health and safety monitoring systems; and

- prompt accident investigation and implementation of remedial actions.

INDICATORS OF A HEALTH AND SAFETY CULTURE

- accident/incident rates;

- sickness and absentee rates;

- resources available for health and safety;

- level of legal and other compliance;

- turnover rates for employees;

- level of complaints;

- selection and management of contractors;

- levels and effectiveness of communication and supervision;

- health and safety management structure; and

- level of insurance premiums.

HUMAN FACTORS

1 in 10 near misses = 1 accident

90% accidents due to human error – 70% poor management

Human factors are affected by the:

- Organisation

- Job

- Personal factors.

ORGANISATION

- must have a positive health and safety culture;

- manage health and safety by providing leadership and involvement of senior managers;

- motivate the workforce to improve health and safety performance; and

- measure health and safety performance.

JOB

- recognise possibility of human error;

- good ergonomics, equipment design and layout of workstation;

- clear job descriptions;

- safe systems of work and operating procedures;

- job rotation and regular breaks;

- provision of correct tools; and

- effective training schedule and good communication.

PERSONAL FACTORS

The three common psychological factors are:

- Attitude – tendency to behave in a particular way in a given situation, influenced by social background and peer pressure

- Motivation – the driving force behind the way a person acts or is stimulated to act

- Perception – the way in which a person believes or understands information supplied or a situation.

Other related factors:

- self-interest – e.g. effect of bonus systems;

- position in the team;

■ acknowledgement by management of good work/initiatives;

■ hearing and/or memory loss;

■ experience and competence;

■ age, personality, attitude, language problems;

■ training undertaken and information given;

■ effect of shift working – e.g. night working; and

■ health (physical and mental).

Human errors may be:	Violations may be:
1 Slips – failure to carry out the correct actions of a task	1 Routine – the breaking of a safety rule or procedure is the normal way of working
2 Lapses – failure to carry out particular actions that form part of a working procedure	2 Situational – job pressures at a particular time make rule compliance difficult
3 Mistakes are: • rule-based – a rule or procedure is applied or remembered incorrectly or • knowledge-based – well tried methods or calculation rules are applied incorrectly	3 Exceptional – a safety rule is broken to perform a new task

METHODS OF IMPROVING HEALTH AND SAFETY BEHAVIOUR AT WORK

DEVELOPMENT OF A POSITIVE HEALTH AND SAFETY CULTURE

By:

1 Commitment of management is the most important factor:

● Proactive management

● Promotion by example (e.g. wearing PPE)

2 The promotion of health and safety standards for:

- Selection and design of premises

- Selection and design of plant, processes and substances

- Recruitment of employees and contractors

- Risk assessments and control implementation

- Competence, maintenance and supervision

- Emergency planning and training

- Transportation of the product and its subsequent maintenance and servicing

3 Competence of the workforce, including in health and safety:

- Knowledge and understanding of the work/job

- Capacity to apply knowledge to the particular task

- Awareness of one's limitations.

COMMUNICATION

- Verbal (by mouth) – conversations, telephone

- Written – memos, emails, meeting minutes, data sheets

- Graphic – safety signs, posters, charts.

Use of notice boards

The following types of health and safety information could be displayed on a workplace notice board:

- a copy of the Employers' Liability Insurance Certificate;

- details of first aid arrangements;

- emergency evacuation and fire procedures;

- minutes of the last health and safety committee meeting;

■ details of health and safety targets and performance against them; and

■ health and safety posters and campaign details.

Barriers to effective communication

■ language and dialect

■ acronyms and jargon

■ various physical and mental disabilities

■ attitudes and perception of workers and supervisors.

Types of accident propaganda

■ Statistics

■ Films, DVDs and posters

■ Targets

■ Records.

For safety propaganda to be effective, it must have:

■ a simple understandable message

■ a positive believable message

■ an appealing format that will motivate the reader.

CONSULTATION WITH EMPLOYEES

Difference between informing and consulting:

■ 'informing' is a one-way process involving the provision of relevant information by management to workers; whereas

■ 'consulting' is a two-way process where account is taken of the views of workers before any decision is taken.

The benefits of consultation include:

■ an improved health and safety culture;

■ motivation of staff;

■ reduction in accidents and ill health;

■ improved overall performance of the organisation.

Role of worker representatives includes to:

■ inform the employer of health and safety concerns of the workforce;

■ inform the employer of potential hazards and dangerous occurrences in the workplace;

■ inform the employer of any general matters that affect the health and safety of the workforce; and

■ speak on behalf of the workforce to health and safety inspectors.

Issues on which the employer should consult:

■ new processes or equipment or any changes in them;

■ the appointment arrangements for a health and safety competent person;

■ the results of any risk assessments;

■ the arrangements for the management of health and safety training; and

■ the introduction of new technologies.

SAFETY COMMITTEES

Terms of reference should include the following:

■ the study of accident and notifiable disease statistics to enable reports to be made of recommended remedial actions;

■ the examination of health and safety audits and statutory inspection reports;

■ the consideration of reports from the external enforcement agency;

■ the review of new legislation, Approved Codes of Practice and guidance and their effect on the organisation;

- the monitoring and review of all health and safety training and instruction activities in the organisation;

- the monitoring and review of health and safety publicity and communication throughout the organisation;

- development of safe systems of work and safety procedures;

- reviewing risk assessments;

- considering reports from safety representatives; and

- continuous monitoring of arrangements for health and safety and revising them whenever necessary.

The employer should not disclose information that:

- violates a legal prohibition

- endangers national security

- relates to a specific individual

- could harm the company commercially

- was obtained from legal proceedings.

TYPES OF HEALTH AND SAFETY TRAINING

- Induction – at recruitment

- Job-specific

- Supervisory and management

- Specialist (e.g. first aid)

- Refresher or reinforcement.

Internal influences on health and safety culture	External influences on health and safety culture
Competence	Expectations of society
Management commitment	Legislation and enforcement
Production/service demands	Insurance companies
Communication	Trade unions
Employee representation	State of the economy
	Commercial stakeholders

EMERGENCY PROCEDURES

EXAMPLES OF TYPES OF EMERGENCIES

- fire
- explosions, bomb scares
- escape of toxic gases
- major accident.

TYPICAL ELEMENTS OF EMERGENCY PROCEDURES

- Fire notices and fire procedures (including testing)
- Fire drills and evacuation procedures
- Assembly and roll call
- Arrangements for contacting emergency and rescue services
- Provision of information for emergency services
- Internal emergency organisation – including control of spillages and clean-up arrangements
- Media and publicity arrangements
- Business continuity arrangements.

PROVISION OF FIRST AID

MAIN FUNCTIONS OF FIRST AID TREATMENT

- preservation of life and/or minimisation of the consequences of serious injury until medical help is available

- treatment of minor injuries not needing medical attention.

MAIN FIRST AID REQUIREMENTS

- qualified first aiders

- adequate facilities and equipment to administer first aid

- an assessment of required first aid cover and requirements

- an appointed person available to assist first aiders.

BASIC FIRST AID PROVISION (INCLUDING NUMBER OF FIRST AIDERS)

depends on:

- number of workers;

- the hazards and risks in the workplace;

- accident record and types of injuries;

- proximity to emergency medical services; and

- working patterns (shift work).

First aider requires first aid training and the scope of the first aid course depends on the level of risk in the workplace. Any training should be given by a recognised training provider.

Appointed person has some other first aid experience/qualification and can deputise during the absence of the trained first aider.

First aid box – contents depend on particular workplace needs but should be checked regularly and not contain medicines.

Health and Safety Management Systems – Planning

LEARNING OUTCOMES

Explain the importance of planning in the context of health and safety management systems

Explain the principles and practice of risk assessment

Explain the general principles of control and a basic hierarchy of risk reduction measures

Identify the key sources of health and safety information

Explain what factors should be considered when developing and implementing a safe system of work

Explain the role and function of a permit-to-work system

KEY REVISION POINTS

The meaning of 'suitable and sufficient' ☐

The forms and objectives of risk assessment ☐

Details of accident categories and types of health risks ☐

The risk assessment process and its management ☐

Groups that require a special risk assessment ☐

The principles of prevention and the hierarchy of risk control ☐

The development and application of safe systems of work
and permits to work for various activities, including those
in confined spaces ☐

Hazards and controls for confined spaces and machinery
maintenance work ☐

THE IMPORTANCE OF PLANNING

The health and safety planning process begins with finding:

■ the correct information about the existing management system for health
and safety;

■ suitable benchmarks against which to make comparisons; and

■ competent people to carry out the analysis and make sensible
judgements.

Further judgement may be necessary to establish if the system is:

■ adequate for the organisation and the range of hazards/risks;

■ working as intended and achieving the right objectives; and

■ delivering cost-effective and proportionate risk control in the workplace.

SETTING HEALTH AND SAFETY OBJECTIVES

The standard of the health and safety objectives of an organisation will depend upon:

- the seniority of the person setting the objectives;

- the formal documentation associated with the objectives and their relevance at each functional level in the organisation;

- the incorporation of relevant legal, technological options and good practice guidance;

- the identification and assessment of all significant hazards and risks within the organisation;

- the success with which they integrate with financial, operational and business requirements; and

- the incorporation of the views of employees, stakeholders and other interested parties.

The objectives must be **specific, measurable, achievable** and with **realistic timescales (SMART)** and kept up to date with any changes in legislation.

PRINCIPLES AND PRACTICE OF RISK ASSESSMENT

Suitable and sufficient means:

- identify significant risks only;

- identify measures required to comply with legislation; and

- remain appropriate and valid over a reasonable period of time.

Hazard – the potential to cause harm

Risk – the likelihood to cause harm

Residual risk – the risk remaining after some controls are in place.

FORMS OF RISK ASSESSMENT

- Quantitative – calculated from risk = likelihood x severity

- Qualitative – descriptor (high, medium or low) used to describe timetable for remedial action

- Generic – covers similar activities or work equipment.

HEALTH RISKS

- Chemical – exhaust fumes, paint solvents, asbestos

- Biological – legionella, other pathogens, hepatitis

- Physical – noise vibration, radiation

- Psychological – stress, violence

- (Ergonomic – musculoskeletal disorders).

RISK ASSESSMENT PROCESS

- Hazard identification – Step 1 of UK HSE's five steps

- Persons at risk – Step 2 of UK HSE's five steps

  - employees, agency/temporary workers, contractors, shift workers

  - members of the public – visitors, customers, patients, students, children, elderly

  - special groups – young persons, expectant or nursing mothers, workers with a disability, lone workers

- Evaluation of risk level (residual risk) – Step 3 of UK HSE's five steps

  - high, medium and low (defined qualitatively or quantitatively)

  - both occupational and organisational risk levels need to be considered

- Detail risk controls (existing and additional) – Step 3 of UK HSE's five steps

  - the prioritisation of risk control is important

  - risks can be reduced at the design stage by using the principles of prevention

  - risks can be controlled by using the hierarchy of risk control (see Element 6)

- Record of risk assessment findings – Step 4 of UK HSE's five steps

- Monitor and review – Step 5 of UK HSE's five steps.

Regular reviews required but need to be more frequent if:

- new legislation introduced

- new information available on substances or process

- changes to the workforce – introduction of trainees

- an accident has occurred.

RISK ASSESSMENT TEAM REQUIREMENTS

- all need training in risk assessment

- leader should have health and safety experience

- all need to be competent to assess risks in area under examination

- all need to know their own limitations

- include local line manager in the team

- at least one team member with report writing skills.

SPECIAL CASES

Young persons

- are under 18 years

- subject to peer pressure and are inexperienced

- are eager to please
- appropriate level and approach of subject matter in training sessions.

A special risk assessment must be made to include details of:

- the work activity;
- any prohibited processes or equipment;
- the health and safety training provided; and
- the supervision arrangements.

Expectant and nursing mothers

There are restrictions on the type of work that can be undertaken.

Risks include:

- manual handling;
- chemical and biological agents;
- ionising radiation;
- passive smoking;
- lack of rest room facilities;
- temperature variations;
- prolonged standing or sitting; and
- stress and violence to staff.

Workers with a disability

- emergency arrangements including 'raising the alarm'; and
- adequate wheelchair access to fire exit.

Lone workers

- special risk assessment (including violence);
- must be fit to work alone;

- special training should be given;

- possible to handle all equipment and substances alone;

- periodic visits by supervisor;

- regular mobile phone contact with base;

- first aid arrangements; and

- emergency arrangements.

GENERAL PRINCIPLES OF CONTROL

The planning and implementing section of the health and safety management system is based on risk assessment and concerns all the actions taken to control or eliminate hazards and reduce risks. The principles of prevention are used when equipment or processes are being designed or selected. The hierarchy of risk control enables risk to be further controlled.

PRINCIPLES OF PREVENTION

- avoid risks

- evaluate risks which cannot be avoided

- adapt work to the individual

- adapt to technical changes

- replace dangerous items with less dangerous items

- develop an overall prevention policy

- give priority to collective measures (Safe Place strategy)

- give instructions to employees (Safe Person strategy).

HIERARCHY OF RISK CONTROL

1 elimination
2 substitution

3 changing work methods/patterns

4 reduced time exposure

5 engineering controls (isolation, insulation and ventilation)

6 good housekeeping

7 safe systems of work

8 training and information

9 personal protective equipment

10 welfare

11 monitoring and supervision

12 review.

SOURCES OF INFORMATION ON HEALTH AND SAFETY

Internal sources	External sources
• accident and ill-health records and investigation reports	• health and safety legislation
• absentee records	• HSE publications, such as approved codes of practice, guidance documents, leaflets, journals, books and their website
• inspection and audit reports undertaken by the organisation and by external organisations such as the HSE	• international (e.g. ILO), European and British standards
• maintenance, risk assessment (including COSHH) and training records	• health and safety magazines and journals
• documents which provide information to workers	• information published by trade associations, employer organisations and trade unions
• any equipment examination or test reports	• specialist technical and legal publications
	• information and data from manufacturers and suppliers
	• the internet and encyclopaedias

SAFETY SIGNS

- Red – prohibition – round (e.g. no smoking)
- Yellow – warning – triangular (e.g. wet floor)
- Blue – mandatory – round (e.g. ear defenders must be worn)
- Green – safe condition – square or rectangular (e.g. first aid).

DEVELOPMENT AND IMPLEMENTATION OF SAFE SYSTEMS OF WORK

EMPLOYER'S DUTY

ILO recommendation R164 requires employers to:

- provide and maintain workplaces, machinery and equipment;
- use work methods which are as safe and without risk to health as is reasonably practicable; and
- give necessary instructions and training, taking account of the functions and capacities of different categories of workers.

ROLE OF COMPETENT PERSONS

A competent person or safety adviser should:

- assist managers to draw up guidelines for safe systems of work;
- prepare suitable documentation; and
- advise management on the adequacy of the safe systems produced.

ROLE OF MANAGERS

- provide safe systems of work;
- ensure that employees are adequately trained in a specific safe system of work and are competent to carry out the work safely; and
- provide sufficient supervision to ensure that the system of work is followed and the work is carried out safely.

EMPLOYEE INVOLVEMENT

Includes:

- consultation with those employees who will be exposed to the risks, either directly or through their representatives;

- discussion of the proposed system with those who will have to work under it and supervise it; and

- understanding that employees have a responsibility to follow the safe system of work.

STEPS IN THE DEVELOPMENT OF A SAFE SYSTEM OF WORK

1 Assess the task (complexity, accident records etc.)
2 Identify the significant hazards associated with the task
3 Define safe methods for performing the task (including emergency procedures) – document the methods if required
4 Implement the safe system of work (written safe system to be signed off)
5 Monitor the safe system of work and review it if necessary
6 Train the workforce in safe procedures and enter on training record.

Safe systems of work are particularly important for:

- maintenance work

- contractors

- emergency procedures

- lone working

- vehicle operations

- cleaning operations.

Method statements are formal written safe systems of work and are often used in construction work.

The safe system of work should be based on a thorough analysis of the job or operation to be covered by the system. After the introduction of a safe system of work the following controls are required:

■ Engineering or process controls (e.g. guarding)

■ Documented procedures

■ Behavioural controls requiring a certain standard of behaviour from individuals.

Procedures for lone workers may include:

■ periodic visits from the supervisor;

■ regular voice contact;

■ automatic warning devices to alert others of problems;

■ checks that the lone worker has returned safely;

■ special arrangements for first aid to deal with minor injuries; and

■ emergency arrangements.

When travelling abroad, ensure that:

■ potential health risks have been addressed;

■ required documentation (e.g. visas) has been obtained;

■ adequate insurance cover has been obtained;

■ any warnings from Embassies etc. concerning unrest have been heeded; and

■ minimum baggage containing few valuables is taken.

ROLE AND FUNCTION OF A PERMIT-TO-WORK

Permits to work are:

- **formal** safe systems of work
- required to be signed on/off by a responsible person
- often require equipment to be locked on/off by a responsible person.

To be used whenever there is a high risk of serious injury, such as:

- confined spaces
- live electricity (particularly high voltage)
- hot working (e.g. welding work)
- some machinery maintenance work.

Typical responsible persons are:

- site manager
- senior authorised person – often the chief engineer
- authorised person – issues permit
- competent person – receives permit
- operatives – supervised by a competent person
- specialists (e.g. electrical engineer)
- engineers – usually responsible for the work
- contractors.

CONFINED SPACES

Examples include underground chamber, silo, trench, sewer, tunnel.

Hazards are:

- lack of oxygen and asphyxiation
- poor ventilation

- presence of fumes

- poor means of access and escape

- drowning

- claustrophobia

- electrical equipment (needs to be flameproof)

- presence of dust (e.g. silos)

- heat and high temperatures

- fire and/or explosion

- poor or artificial lighting.

Controls include:

- permit-to-work

- risk assessment

- training and information for all workers entering the confined space

- emergency arrangements in place

- emergency training

- no entry for unauthorised persons.

MACHINERY MAINTENANCE WORK

Hazards are:

- no perceived risk

- no safe system of work

- poor communications

- failure to brief contractors

- lack of familiarity

- poor design.

Control of hazards requires:

- effective planning

- a written safe system of work/permit-to-work

- a risk assessment to assess, control and reduce risks

- monitoring to ensure that the system of work and controls are used

- an effective training programme for all involved in the work.

Health and Safety Management Systems – Measuring, Audit and Review

1.5

LEARNING OUTCOMES

Outline the principles, purpose and role of active and reactive monitoring ☐

Explain the purpose of, and procedures for, health and safety auditing ☐

Explain the purpose of, and procedures for, investigating incidents (accidents, cases of work-related ill health and other occurrences) ☐

Describe the legal and organisational requirements for recording and reporting incidents ☐

Explain the purpose of, and procedures for, regular reviews of health and safety performance ☐

 KEY REVISION POINTS

The reasons for measuring and monitoring health and safety performance ☐

The role of standards in the monitoring process ☐

The differences between reactive and active (proactive) monitoring ☐

The importance of clear unambiguous report writing ☐

The role of performance review and audit ☐

The advantages and disadvantages of internal and external audit ☐

The reasons and legal requirements for recording and reporting incidents and accidents ☐

Basic accident investigation procedures and the different types of accident and incident ☐

Immediate and underlying causes of incidents ☐

Issues concerning insurance and compensation claims ☐

WORKPLACE INSPECTIONS

Detailed check, often using a checklist, of the whole workplace and should cover:

- ▩ the premises (e.g. fire precautions, access/egress, housekeeping);

- ▩ the plant, equipment and substances (e.g. machine guarding, tools, ventilation);

- ▩ the procedures in place (e.g. safe systems of work, risk assessments, use of PPE); and

- ▩ the workforce (e.g. training, information, supervision, health surveillance).

ACTIVE AND REACTIVE MONITORING

Reactive monitoring (taking action after a problem occurs) involves	Proactive (or active) monitoring (taking action before problems occur) involves
• review of accident and ill-health reports – often to check that remedial advice has been actioned or ascertain trends and hot spots • review of procedures following dangerous occurrences, other property damage and near misses • review of compensation claims	• the active monitoring of the workplace for unsafe conditions • the direct observation of workers for unsafe acts • meeting with management and workers to discover any problems
• review of complaints from the workforce and members of the public • review of procedures following enforcement reports and notices • review of risk assessments following the discovery of additional hazards	• checking documents, such as maintenance records, near miss reports, insurance reports • undertaking workplace inspections, sampling, surveys, tours and audits

Other issues with inspections

▨ the competence of the observers;

▨ the frequency of inspections;

▨ the response to the inspection report; and

▨ the use of objective inspection standards.

Safety sampling – checking for safety defects in a selected area of the work-place (e.g. all fire extinguishers).

Safety surveys – a detailed inspection of a particular workplace activity throughout an organisation (e.g. manual handling).

Safety tours – an unscheduled brief inspection of a work area in the workplace by a team led by a senior manager.

HEALTH AND SAFETY AUDITING

Auditing – the independent collection of information on the efficiency, effectiveness and reliability of the whole health and safety management system measured against specific standards. It will check that the following are in place:

▪ appropriate management arrangements;

▪ adequate risk control systems exist and implemented;

▪ appropriate workplace precautions; and

▪ appropriate documentation and records.

Audits should take place at regular intervals.

OTHER ISSUES WITH AUDITS

▪ External audits:

- are independent of organisation

- are competent

- are familiar with external benchmarks

- are more expensive than internal audits

- do not know the organisation

- require more information than internal audits

- can offer bland reports.

▪ Internal audits:

- are not independent of organisation and partiality may be questioned

- usually require audit training

- are less expensive than external audits

- know the organisation well, particularly critical areas

- can spread good practice around the organisation
- may be unaware of external benchmarks.

INVESTIGATING INCIDENTS

Purpose of accident/incident investigation is to:

- eliminate the cause and future occurrences;
- determine the direct and indirect causes of the accident/incident;
- define any corrective and/or preventative actions;
- identify any deficiencies in risk controls, the health and safety management system and/or procedures;
- ensure that essential information is available in the event of a compensation claim; and
- ensure that all legal requirements are being met.

Benefits of accident/incident investigation:

- prevention of a recurrence;
- prevention of future business losses;
- prevention of future increased insurance premiums and costs of criminal and civil actions; and
- improve employee morale and organisation reputation.

ACCIDENT CAUSATION

Direct or immediate include	Indirect, root or underlying include
• unsafe acts by individuals due to poor behaviour or a lack of training, supervision, information or competence or a failure to wear PPE	• poor machine maintenance and/or start-up procedures • management and social pressures • financial restrictions • lack of management commitment to health and safety
• unsafe conditions, such as inadequate guarding, inadequate procedures, hazardous substances, ergonomic and/or environmental factors • fire or explosive hazards	• poor or lack of health and safety policy and standards • poor workplace health and safety culture • workplace and trade customs and attitudes

ELEMENTS OF AN INVESTIGATION

- interview relevant operatives, managers, supervisors

- obtain detailed plans and/or photographs of scene

- check all relevant records (working procedures, maintenance, training, risk assessments)

- interview all witnesses

- possibly arrange for equipment and/or substances to be independently tested

- produce a concise report for management that includes

    - background to the accident

    - possible causes of the accident (direct and indirect)

    - relevant health and safety legislation, guidance and standards

- recommendations (including any remedial actions)
- any additional training or follow-up requirements

- undertake a post-accident risk assessment.

RECORDING AND REPORTING INCIDENTS

Most organisations will want to collect data on:

- all injury accidents;

- cases of ill health;

- sickness absence;

- damage to property, personal effects and work in progress; and

- incidents with the potential to cause serious injury, ill health or damage.

There are several ways in which data can be analysed and presented. The most common ways are:

- by causation;

- by the nature of the injury, such as cuts, abrasions, asphyxiation and amputations;

- by the part of the body affected, such as hands, arms, feet, lower leg, upper leg, head, eyes and back;

- by time of day;

- by occupation or location of the job; and

- by physical agency involved, such as machines, means of transport and substances.

INCIDENT REPORTING REQUIREMENTS

■ the organisational accident/incident report form

■ accident book

■ national accident reporting procedures for external agencies.

The ILO Code of Practice requires that occupational accidents are classified as:

■ Occupational accidents resulting in death;

■ Occupational non-fatal accidents with at least three consecutive days of incapacity excluding the day of the accident;

■ Commuting accidents;

■ Occupational diseases as defined in national laws; and

■ Dangerous occurrences as defined by national laws.

The employer or a responsible person is required to notify the authorities according to national laws and regulations:

■ by the quickest possible means immediately after a fatal occupational accident; or

■ within a prescribed time for other occupational accidents and occupational diseases.

COMPENSATION AND INSURANCE ISSUES

■ Importance of documentation and records

■ Need for pre- and post-accident risk assessments.

REVIEW OF HEALTH AND SAFETY PERFORMANCE

Performance review is the final stage of the management process and reviews:

■ all monitoring, inspection and audit reports;

- the adequacy of the health and safety management system and performance standards against external benchmarks;

- whether new legislation or guidance has been applied;

- whether the health and safety policy objectives have been met or need modification to ensure continuous improvement; and

- there has been adequate feedback to/from managers.

Reviews should aim to include:

- evaluation of compliance with legal and organisational requirements;

- incident data, recommendations and action plans from investigations;

- inspections, surveys, tours and sampling;

- absences and sickness records and their analysis;

- any reports on quality assurance or environmental protection;

- audit results and implementation;

- monitoring of data, reports and records;

- communications from enforcing authorities and insurers;

- any developments in legal requirements or best practice within the industry;

- changed circumstances or processes;

- benchmarking with other similar organisations;

- complaints from neighbours, customers and the public;

- effectiveness of consultation and internal communications;

- whether health and safety objectives have been met; and

- whether actions from previous reviews have been completed.

The Board should review health and safety performance at least once a year. The review process should:

■ examine whether the health and safety policy requires revision;

■ examine whether risk management and other health and safety issues have been effectively reported to the Board;

■ report health and safety shortcomings;

■ address any weaknesses with effective measures and develop a system to monitor their implementation; and

■ undertake immediate reviews in the light of major shortcomings or events.

Control of International Workplace Risks

Workplace Hazards and Risk Control

LEARNING OUTCOMES

Outline common health, welfare and work environment requirements in the workplace ▪

Explain the risk factors and appropriate controls for violence at work ▪

Explain the effects of substance misuse on health and safety at work and control measures to reduce such risks ▪

Explain the hazards and control measures for the safe movement of people in the workplace ▪

Outline the hazards and control measures for safe construction and demolition work ▪

Explain the hazards and control measures for safe working at height ▪

Explain the hazards of, and control measures for, excavations ▪

KEY REVISION POINTS

Welfare and work environment requirements ☐

The causes and prevention of workplace violence ☐

The hazards and control strategies for pedestrian safety ☐

The hazards and controls in construction work ☐

The hazards and controls involved with working at height, including roof work ☐

Access equipment, including scaffolds, ladders and MEWPs ☐

Emergency rescue procedures ☐

The hazards and controls required for excavation work ☐

HEALTH, WELFARE AND WORK ENVIRONMENT REQUIREMENTS

WELFARE ISSUES

- ▪ sanitary conveniences and washing facilities

- ▪ drinking water

- ▪ clothing and changing facilities

- ▪ rest and eating facilities.

WORKPLACE ENVIRONMENT

- ▪ ventilation

- ▪ heating and temperature (particularly extremes)

- ▪ lighting

  - ● natural light

  - ● stroboscopic effects

- special local lighting
- structural aspects (shadows)
- atmospheric dust
- heating effect of lighting
- lamp and window cleaning
- emergency lighting

■ workstations and seating

■ floors, stairways and traffic routes

■ translucent or transparent doors should be constructed with safety glass and properly marked to warn pedestrians of their presence

■ adequate arrangements in place to ensure the safe cleaning of windows and skylights

■ adequate provisions for the needs of disabled workers.

EFFECTS OF EXTREME TEMPERATURE

	Health effects	Preventative measures
High temperatures	Heart strain Heat stroke	Supply of drinking water Supply of ventilated air Regular health surveillance Information and training on the hazards
Low temperatures	Frostbite Loss of limbs	Supply of thick, warm (thermal) clothing Provision of hot drinks and external heating Regular health surveillance Information and training on the hazards

VIOLENCE AT WORK

Defined as any incident in which a person is abused, threatened or assaulted in circumstances related to their work. High-risk occupations include health and social services, police and fire fighters, lone and night workers and benefit services.

ACTION PLAN

1   find out if there is a problem
2   decide on what action to take:

  i.   quality of service provided

  ii.   design of the operating environment

  iii. type of equipment used

  iv.   designing the job

3   take the appropriate action
4   check that the action is effective.

TYPE OF SECURITY EQUIPMENT USED

- access control (swipe cards and simple coded security locks);

- closed circuit television;

- alarms – there are three main types:

  - intruder alarms

  - panic alarms

  - personal alarms

- radios and pagers (particularly for lone workers); and

- mobile phones.

SUBSTANCE MISUSE AT WORK

RISKS TO HEALTH AND SAFETY OF DRUGS AND ALCOHOL

- reduce productivity;

- increase absenteeism;

- increase accidents at work; and, in some cases,

- endanger the public.

Alcohol abuse is a considerable problem when vehicle driving is part of the job, especially if driving is required on public roads.

CONTROLS

For alcohol abuse:

- during working hours, there should be no drinking

- drinking during break periods should be discouraged

- induction training should stress this policy

- managers should enforce the policy

- posters can also help to communicate the message

- the availability of a confidential counselling service.

Drug abuse presents similar problems to those found with alcohol abuse:

- absenteeism

- reduced productivity

- an increase in the risk of accidents

- impairments in cognition, perception and motor skills.

SIGNS OF A SUBSTANCE MISUSE PROBLEM

- sudden mood changes

- unusual irritability or aggression

- a tendency to become confused

- abnormal fluctuations in concentration and energy

- impaired job performance

- poor time-keeping

- increased short-term sickness leave

- a deterioration in relationships with colleagues, customers or management

- dishonesty and theft (arising from the need to maintain an expensive habit).

A policy on drug abuse can be established by:

1 Investigation of the size of the problem
2 **Planning actions** – develop an awareness programme for all staff and a special training programme for managers and supervisors
3 **Taking action** by producing a written policy to include:

    i   details of the safeguards to employees

    ii   the confidentiality given to anyone with a drug problem

    iii the circumstances in which disciplinary and/or reported action will be taken

4 **Monitoring the policy** by checking for positive changes in the measures made during the initial investigation (improvements in the rates of sickness and accidents).

MOVEMENT OF PEOPLE

Hazards to pedestrians include:

- slips, trips and falls on the same level

- falls from height

- collisions with moving vehicles

- being struck by moving, falling or flying objects

- striking against fixed or stationary objects.

Slip hazards are caused by:

- wet or dusty floors
- the spillage of wet or dry substances – oil, water, plastic pellets
- loose mats on slippery floors
- wet and/or icy weather conditions
- unsuitable footwear, floor coverings or sloping floors.

Trip hazards are caused by:

- loose floorboards or carpets
- obstructions, low walls, low fixtures on the floor
- cables or trailing leads to portable electrical hand tools and other electrical appliances across walkways
- raised telephone and electrical sockets
- rugs and mats – particularly when worn or placed on a polished surface
- poor housekeeping – obstacles left on walkways, rubbish not removed regularly
- poor lighting levels – particularly near steps or other changes in level
- sloping or uneven floors – particularly where there is poor lighting or no handrails
- unsuitable footwear – shoes with a slippery sole or lack of ankle support.

CONTROL STRATEGIES FOR PEDESTRIANS

- risk assessment to identify suitable controls
- slip-resistant surfaces and reflective edges to stairs and kerbs
- spillage control and drainage
- designated walkways
- fencing and guarding, particularly on stairways

- use of warning signs

- sound storage racking that is inspected and maintained regularly

- maintenance of a safe workplace

- cleaning and good housekeeping procedures

- access and egress

- environmental considerations (heating, lighting, noise and dust)

- special provision for disabled people

- high visibility clothing, appropriate footwear, personal protective equipment (safety harnesses)

- information, instruction, training and supervision.

CONSTRUCTION HAZARDS AND CONTROLS

The scope of construction includes:

- general building work – domestic, commercial or industrial and may be:

  - new building work, such as a building extension, or

  - the refurbishment, renovation, maintenance or repair of existing buildings

- civil engineering projects involving road and bridge building, water supply and sewage schemes and river and canal work;

- the use of woodworking workshops together with woodworking machines and their associated hazards;

- electrical installation and plumbing work;

- painting and decorating; and

- work in confined spaces such as excavations and underground chambers.

HAZARDS AND CONTROLS

1 Safe place of work

- secure and locked gates with appropriate notices posted;

- a secure and undamaged perimeter fence with appropriate notices posted;

- all ladders either stored securely or boarded across their rungs;

- all excavations covered;

- all mobile plant immobilised, where practicable, and services isolated;

- safe stacking of materials;

- secure storage of all inflammable and hazardous substances;

- visits to local schools to explain the dangers present on a construction site;

- if unauthorised entry persists, then security patrols and closed circuit television may need to be considered.

2 Protection against falls

When working at height, the following hierarchy of control should be followed:

1 Avoid working at height if possible

2 Use an existing safe place of work

3 Provide work equipment to prevent falls

4 Mitigate distance and consequences of a fall

5 Instruction, training and supervision.

Following a suitable risk assessment, the following hierarchy of measures should be considered:

1 avoid working at height, If possible;

2 the provision of a properly constructed working platform, complete with toe boards and guard rails;

3   if this is not practicable or where the work is of short duration, suspension equipment should be used, and only when this is impracticable;

4   collective fall arrest equipment (air bags or safety nets) may be used;

5   where this is not practicable, individual fall restrainers (safety harnesses) should be used;

6   only when none of the above measures are practicable, should ladders or stepladders be considered.

3   Fragile roofs

Particular hazards are:

- fragile roofing materials – more brittle with age and exposure to sunlight;

- exposed edges;

- unsafe access equipment;

- falls from girders, ridges or purlins;

- overhead services and obstructions;

- the use of equipment such as gas cylinders and bitumen boilers; and

- manual handling hazards.

Controls include:

- suitable means of access such as scaffolding, ladders and crawling boards;

- suitable barriers, guard rails or covers where people work near to fragile materials and roof lights; and

- suitable warning signs indicating that a roof is fragile, at ground level.

4 **Protection from falling objects:**

- the use of covered walkways or suitable netting to catch falling debris;
- waste material should be brought to ground level by the use of chutes or hoists;
- minimal quantities of building materials should be stored on working platforms; and
- hard hats must be given to all employees whenever there is a risk of head injury from falling objects. Self-employed workers must supply their own head protection. Visitors to construction sites should be supplied with head protection and mandatory head protection signs displayed around the site.

5 **Demolition controls include:**

- prior to work, a full site investigation;
- risk assessments (two required – project designer and contractor);
- a method statement;
- waste disposal arrangements;
- piecemeal and deliberate controlled collapse; and
- training and information to all workers.

6 **Prevention of drowning**

Arrangements in place to prevent people falling into the water and ensure that rescue equipment is available at all times.

7 **Vehicles and traffic routes**

- Types of construction vehicles include fork-lift trucks, dumper trucks, excavators and cement mixers.
- All vehicles must be well maintained and only driven by trained persons.
- Traffic routes, loading and storage areas need to be well designed with enforced speed limits, good visibility and the separation of vehicles and pedestrians being considered.

- The use of one-way systems and separate site access gates for vehicles and pedestrians may be required.

- The safety of members of the public must be considered, particularly where vehicles cross public footpaths.

8 Fire and other emergencies

Emergency procedures required for fire, explosions, flooding or structural collapse and should include:

- the location of fire points and assembly points;

- accident reporting and investigation; and

- rescue from excavations and confined spaces.

9 Welfare facilities to include:

- sanitary and washing facilities (including showers if necessary) with an adequate supply of drinking water;

- accommodation for the changing and storage of clothes;

- rest facilities for break times;

- adequate first aid provision (an accident book); and

- protective clothing against adverse weather conditions.

10 Electricity

- only 110 V equipment should be used on site;

- if mains electricity is used, then residual current devices should be used with all equipment;

- where workers or tall vehicles are working near or under overhead power lines, either the power should be turned off or 'goal posts' or taped markers used to prevent contact with the lines; and

- similarly, underground supply lines should be located and marked before digging takes place.

11 **Noise**

- noisy machinery should be fitted with silencers;

- ear defenders issued when working with noisy machinery; and

- when machinery is used in a workshop (such as woodworking machines), a noise survey should be undertaken.

12 **Additional health hazards include:**

- dust – including asbestos;

- vibration;

- cement dust and wet cement;

- wood dust;

- solvents, cleaners and paints;

- tetanus and other biological agents; and

- manual handling.

13 **Waste disposal**

- the disposal of waste must be properly managed;

- waste skips must not be overloaded or used for inappropriate waste.

MANAGEMENT OF CONSTRUCTION ACTIVITIES

An ILO code of practice 'Safety and Health in Construction' provides guidance in the implementation of the provisions of the Safety and Health in Construction Convention, 1988 (No. 167) and the Safety and Health in Construction Recommendation, 1988 (No. 175). This covers:

- General duties of employers;

- The duties of co-operation and co-ordination when several employers operate on a construction site;

- The rights and duties of construction workers; and

- The general duties of construction designers, engineers and architects.

GENERAL DUTIES OF DESIGNERS, ENGINEERS, ARCHITECTS AND CLIENTS

1 Those concerned with the design and planning of a construction project should receive training in safety and health and should integrate the safety and health of the construction workers into the design and planning process in accordance with national laws, regulations and practice.

2 Care should be exercised by engineers, architects and other professional persons not to include anything in the design which would necessitate the use of dangerous structural or other procedures or materials hazardous to health or safety which could be avoided by design modifications or by substitute materials.

3 Those designing buildings, structures or other construction projects should take into account the safety problems associated with subsequent maintenance and upkeep where maintenance and upkeep would involve special hazards.

4 Facilities should be included in the design for such work to be performed with the minimum risk.

The code also covers the general duties of clients of construction projects:

1 Clients should:

   i   co-ordinate or nominate a competent person to co-ordinate all activities relating to safety and health on their construction projects;

   ii   inform all contractors on the project of special risks to health and safety of which the clients are or should be aware; and

   iii require those submitting tenders to make provision for the cost of safety and health measures during the construction process.

2 In estimating the periods for completion of work stages and overall completion of the project, clients should take account of safety and health requirements during the construction process.

SELECTION AND CONTROL OF CONTRACTORS

The selection of contractors is covered in Element 3 (3.1) of IGC1. On being selected, contractors should be expected to:

- familiarise themselves with the health and safety aspects of the project that may affect their work; and

- co-operate with the main site contractor.

On arrival at the site, sub-contractors should ensure that:

- they report to the Site Office and the Site Manager;

- they abide by any site rules, particularly in respect of PPE;

- the performance of their work does not place others at risk;

- they are familiar with the first aid and accident reporting arrangements on the site;

- they are familiar with all emergency procedures on the site;

- any materials brought onto the site are safely handled, stored and disposed;

- they adopt adequate fire precaution and prevention measures when using equipment which could cause fires;

- they minimise noise and vibration produced by their equipment and activities;

- any ladders, scaffolds and other means of access are erected in conformance with good working practice;

- any welding or burning equipment brought to the site is in a safe operating condition and used safely with a suitable fire extinguisher to hand;

- any lifting equipment brought onto the site complies with any relevant national legislation;

- all electrical equipment complies with the local statutory requirements;

- connections to the electricity supply are from a point specified by the main contractor and are by proper cables and connectors. For outside construction work, only 110 V equipment should be used;

- any restricted access to areas on the site is observed;

- welfare facilities provided on site are treated with respect; and

- any vehicles brought onto the site observe any speed, condition or parking restriction.

WORKING AT HEIGHT

Work at height includes all work activities where there is a need to control a risk of falling a distance liable to cause personal injury. They would therefore include:

- working on a scaffold or from a mobile elevated work platform (MEWP);

- sheeting a lorry or dipping a road tanker;

- tree surgery and other forestry work at height;

- using cradles or rope for access to a building or other structure like a ship under repair;

- climbing permanent structures like a gantry or telephone pole;

- working near an excavation area or cellar opening if a person could fall into it and be injured;

- painting or pasting and erecting bill posters at height;

- work on staging or trestles, for example, for filming or events;

- using a ladder/stepladder or kick stool for shelf filling, window cleaning and the like; and

- working in a mine shaft or chimney.

The following hierarchy of control should be used when work at height is under consideration:

1 avoid working at height, If possible;
2 the provision of a properly constructed working platform, complete with toe boards and guard rails;
3 if this is not practicable or where the work is of short duration, suspension equipment should be used and only when this is impracticable;
4 collective fall arrest equipment (air bags or safety nets) may be used;
5 where this is not practicable, individual fall restrainers (safety harnesses) should be used; and
6 only when none of the above measures are practicable, should ladders or stepladders be considered.

FRAGILE ROOFS AND SURFACES

A roof should always be treated as fragile until a competent person has ruled otherwise. Fragility can be caused by:

▧ general deterioration of the roof through ageing, neglect and lack of maintenance;

▧ corrosion of cladding and fixings;

▧ quality of the original installation and selection of materials;

▧ thermal and impact damage;

▧ deterioration of the supporting structure; and

▧ weather damage.

Asbestos cement sheets and old roof lights should always be treated as fragile.

Other roof work hazards:

▧ overhead services and obstructions;

▧ the presence of asbestos or other hazardous substances;

▧ the use of equipment such as gas cylinders and bitumen boilers; and

▧ manual handling hazards.

REQUIREMENTS FOR HEAD PROTECTION

The ILO code of practice 'Safety and Health in Construction' recommends that employers should supply head protection (hard hats) to employees whenever there is a risk of head injury from falling objects.

Other requirements:

▧ the employer must ensure that hard hats are properly maintained and replaced when they are damaged;

▧ self-employed workers must supply and maintain their own head protection;

- visitors to construction sites should always be supplied with head protection; and

- mandatory head protection signs displayed around the site.

FALL ARREST EQUIPMENT

The three most common types are:

- safety harnesses

- safety nets

- air bags.

EMERGENCY PROCEDURES (INCLUDING RESCUE)

A suitable emergency and rescue procedure needs to be in place for situations that could be expected where people are working at height.

Factors to be considered:

- the method of rescue must be simple and straightforward;

- any rescue system selected needs to be proportionate to the risk;

- there should not be undue reliance on the emergency services. However, arrangements should be in place to inform the emergency services of any serious incident;

- awareness of suspension trauma;

- rescue teams should be available at all times with a designated leader; and

- instruction and training are essential for the teams, preferably by simulated rescue exercises.

ACCESS EQUIPMENT

LADDERS

The following factors should be considered when using ladders	Certain work should not be attempted using ladders. This includes work where
• aluminium ladders are light but should not be used in high winds or near live electricity • timber ladders need regular inspection for damage and should not be painted as this could hide cracks • ensure that the use of a ladder is the safest means of access for the work to be done and the height to be climbed • the ladder needs to be stable in use with a safe inclination (1 in 4) • the foot of the ladder should be tied to a rigid support • the proximity of live electricity should be checked • there should be at least 1 metre of ladder above the stepping-off point • over-reaching must be eliminated • workers who are to use ladders must be trained in the correct method of use and selection • ladders should be inspected (particularly for damaged or missing rungs) and maintained on a regular basis and they should only be repaired by competent persons	• two hands are required • the work is at an excessive height • the ladder cannot be secured or made stable • the work is of long duration • the work area is very large • the equipment or materials to be used are heavy or bulky • the weather conditions are adverse • there is no protection from vehicles

FIXED SCAFFOLD

Fixed scaffolds are usually independently tied and important considerations are:

- scaffolding must only be erected and dismantled by competent people – any changes to the scaffold must be made by a competent person;

- adequate toe boards, guard rails and intermediate rails must be fitted to prevent people or materials from falling;

- the scaffold must rest on a stable surface; uprights should have base plates and timber sole plates if necessary;

- the scaffold must have safe access and egress;

- work platforms should be fully boarded with no tipping or tripping hazards;

- the scaffold should be sited away from or protected from traffic routes so that it is not damaged by vehicles;

- the scaffold should be properly braced and secured to the building or structure;

- overloading of the scaffold must be avoided;

- the public must be protected at all stages of the work; and

- regular inspections of the scaffold must be made and recorded.

Components of a scaffold:

- Standard

- Ledger

- Guard rail

- Toe boards

- Bracing

- Transom

- Base plate

- Sole board

- Ties

- Working platform.

Prefabricated mobile scaffold towers

The following points must be considered:

- the selection, erection and dismantling of mobile scaffold towers must be undertaken by competent persons;

- maximum height to base ratios must not be exceeded;

- diagonal bracing and stabilisers should be used;

- access ladders must be fitted to the narrowest side of the tower or inside the tower;

- persons should not climb up the frame of the tower;

- all wheels must be locked whist work is in progress;

- all persons must vacate the tower before it is moved;

- the tower working platform must be boarded, fitted with guard rails and toe boards and not overloaded;

- towers must be tied to a rigid structure if exposed to windy weather;

- persons working from a tower must not over-reach or use ladders from the work platform;

- safe distances must be maintained between the tower and overhead power lines both during working operations and when the tower is moved; and

- the tower should be inspected on a regular basis and a report made.

MOBILE ELEVATED WORK PLATFORMS (MEWPs)

The following factors must be considered when using mobile elevated work platforms:

- the mobile elevated work platform must only be operated by trained and competent persons;

- safety harnesses should be worn;

- it must never be moved in the elevated position;

- it must be operated on level and stable ground;

- the tyres must be properly inflated and the wheels immobilised;

- outriggers should be fully extended and locked in position;

- due care must be exercised with overhead power supplies and obstructions; and

- procedures should be in place in the event of machine failure.

ROOF WORK

The main hazards are:

- fragile roofing materials – more brittle with age and exposure to sunlight;

- exposed edges;

- unsafe access equipment;

- fragile roof lights and voids;

- falls from girders, ridges or purlins;

- overhead services and obstructions;

- the use of equipment such as gas cylinders and bitumen boilers; and

- manual handling hazards.

The controls include:

- a risk assessment and a method statement must be completed before any roof work takes place;

- suitable means of access such as scaffolding, ladders and crawling boards;

- suitable barriers, guard rails or covers where people work near to fragile materials and roof lights;

- edge protection on a flat roof is best provided by double guard rails and toe board;

- suitable warning signs indicating that a roof is fragile, at ground level;

- access to the area immediately below the work should be restricted using suitable barriers, netting, safety signs and safety helmets;

- the safe transportation of materials to and from the working area, involving lifts, hoists, manual handling and/or using chutes and covered waste skips; and

- good housekeeping procedures.

A **risk assessment for roof work** should include:

- an assessment of the structural integrity of the roof;

- the methods to be used to repair the roof; and

- an assessment of any hazardous substances to be used.

CAUSES OF ACCIDENTS WITH ACCESS EQUIPMENT

- Design faults

- Over-reaching

- Falls and slips

- Collision with vehicles

- Unsound base support

- Untrained personnel

- Climbing with loads

- Poor maintenance and/or inspection

- Instability

- Adverse wind effects.

INSPECTION AND MAINTENANCE

Equipment for work at height needs regular inspection and maintenance. The requirements for inspection are:

- the name and address of the person for whom the inspection was carried out;

- the location of the work equipment inspected;

- a description of the work equipment inspected;

- the date and time of the inspection;

- details of any matter identified that could give rise to a risk to the health or safety of any person;

- details of any action taken as a result of any matter identified above;

- details of any further action considered necessary; and

- the name and position of the person making the report.

EXCAVATIONS

HAZARDS

- Collapse of sides

- Materials falling on workers

- Falls of people and/or vehicles into excavation

- Influx of groundwater

- Underground services

- Access and egress

- Proximity of waste or stored materials or adjacent structures

- Fumes and health hazards (Weil's disease).

PRECAUTIONS

- Supervision by a competent person

- Hard hats

- Shore sides

- Barrier around top

- Well lit at night

- Identify position of buried services

- Care during the filling in.

INSPECTIONS

- Every 7 days with a report

- Every day:

    - After any event likely to affect the stability of excavation

    - Before start of each shift

    - After an accidental fall of material.

Transport Hazards and Risk Control

LEARNING OUTCOMES

Explain the hazards and control measures for the safe movement of vehicles in the workplace

Outline the factors associated with driving at work that increase the risk of an incident and the control measures to reduce work-related driving risks

 KEY REVISION POINTS

The hazards and control strategy for safe vehicular operations ☐

The management of vehicle movements in the workplace ☐

The management of occupational driving risks in terms of
the driver, vehicle and journey planning ☐

The particular issues in the use of lorries ☐

SAFE MOVEMENT OF VEHICLES IN THE WORKPLACE

Hazards in vehicle operations include:

- collisions between pedestrians and vehicles

- vehicles crushing feet of pedestrians

- people falling from vehicles

- people being struck by objects falling from or attached to vehicles

- people being struck by or ejected from an overturning vehicle

- communication problems between vehicle drivers and employees or members of the public.

Hazards in vehicle operations may be caused by:

- poor working practices, such as the lack of regular vehicular safety checks

- defective maintenance, steering, brakes, tyres and hydraulic hoses

- poor road surfaces and/or poorly drained road surfaces

- overloading of vehicles

- use of unsuitable vehicles to transport people

- inadequate supervision

- poor training or lack of refresher training.

Other more general hazards involving pedestrians and vehicles include:

■ reversing of vehicles, especially inside buildings;

■ roadways too narrow with insufficient safe parking areas;

■ roadways poorly marked out and inappropriate or unfamiliar signs used;

■ too few pedestrian crossing points;

■ the non-separation of pedestrians and vehicles;

■ lack of barriers along roadways;

■ lack of directional and other signs;

■ poor visibility due to load impeding view, blind corners or inadequate vehicular lighting or mirrors;

■ poor environmental factors, such as lighting, dust and noise;

■ ill-defined speed limits and/or speed limits which are not enforced; and

■ vehicles used by untrained and/or unauthorised personnel.

HAZARDS FROM MOBILE WORK EQUIPMENT

Accidents involving mobile work equipment often arise from one or more of the following events:

■ poor maintenance with defective brakes, tyres and steering;

■ poor visibility;

■ operating on rough ground or steep gradients;

■ carrying of passengers without the proper accommodation for them;

■ people being ejected as the vehicle overturns and/or being crushed by it;

■ being struck by a vehicle or an attachment;

■ lack of driver training or experience;

■ poor management procedures, training and supervision;

■ collision with other vehicles;

- overloading of vehicles;

- general vehicle movements and parking;

- dangerous occurrences or other emergency incidents; and

- access and egress from the buildings.

CONTROL STRATEGIES FOR SAFE VEHICLE OPERATIONS

- risk assessment

- designated marked traffic routes with good visibility and well-designed loading and storage areas

- enforced speed limits and/or the fitting of speed governors

- audible warning of approach signals, particularly for reversing

- separation of pedestrians and vehicles

- use of one-way systems

- the issue of suitable high visibility clothing and other personal protective equipment to drivers and pedestrians in vehicle operating areas

- separate site access gates for pedestrians and vehicles

- use of pedestrian crossings on vehicular routes

- mirrors on blind corners

- identification of recognised parking and non-parking areas

- include vehicle safety in employee induction training

- driver training and refresher training given by competent persons

- the provision of driver protection by the use of:

  - seat belts

  - FOPS (fall over protection system)

  - ROPS or TOPS (roll or tip over protection systems)

  - banksmen during reversing operations

- effective vehicular maintenance procedures

- use of daily vehicular inspection checklists

- environmental considerations (road surfaces, gradients and changes in road level, lighting, visibility).

The management of vehicle movements involves:

- a designated management system and a code of practice for all drivers, who must be competent

- an effective line management organisation to ensure supervision of all vehicular activities

- a documented preventative maintenance programme with all work recorded

- regular inspections and, in some cases, thorough examinations by competent persons

- ensure that all vehicles are fitted with suitable driver protection – seat belts, ROPS

- ensure that all vehicles are fitted with reversing warning systems

- suitable fire precautions, particularly in battery charging areas.

DRIVING AT WORK

FACTORS AFFECTING OCCUPATIONAL ROAD RISKS

The driver:

- competency issues including:

    - validity of the driving licence

    - awareness of the company policy on work-related road safety

    - general driving experience

- health and fitness, in particular eyesight

- level of training received.

The vehicle:

- its suitability;

- its condition, particularly seat belts and head restraints;

- its past maintenance schedule;

- the ergonomics of the driving position;

- any load should be stacked safely so that it cannot move during the journey. There must also be satisfactory arrangements for handling the load at either end of the journey; and

- the vehicle breakdown procedure.

The journey:

- route to be undertaken, including the amount of traffic, the time of day when the journey is made and the sufficiency of the time allowed for its completion;

- the type of road being used, whether motorway, minor roads or those involving city centre congestion, and their condition;

- schedules need to take sufficient account of busy periods on roads;

- driving excessive distances without appropriate breaks; and

- adverse weather conditions, such as snow, ice, heavy rain and high winds.

MANAGING OCCUPATIONAL ROAD RISKS

An effective policy to manage occupational road risk should address important and relevant aspects of the Highway Code, the possible enforcement action by the authorities and the following issues:

For the driver

- all drivers must have a current and valid driving licence;

- details of a prospective driver's previous driving experience should be examined and recorded;

- driver training and health;

- regular eyesight tests;

- staff should not drive, or undertake other duties, while taking a course of medicine that might impair their judgement;

- some prescribed and over the counter drugs and medicines can affect driver awareness and speed of reaction. Drivers using such medicines must check with a doctor or pharmacist to ensure that it is safe to drive;

- the monitoring of driver performance;

- the number of recorded incidents;

- all speed limits must be observed;

- the determination and recording of driving hours;

- routine driver safety checks such as those on lights, tyres and wheel fixings;

- drivers and passengers are adequately protected in the event of an incident. Crash helmets and protective clothing for those who ride motorcycles and other two-wheeled vehicles should be of the appropriate colour and standard;

- drivers of Heavy Goods Vehicles (HGVs) must have the appropriate driving licence and medical certificate; and

- disciplinary action taken in the event of non-compliance with the agreed in-house rules and standards.

For the vehicle

- all vehicles must carry comprehensive insurance for use at work;

- all vehicles should be fit for purpose;

- new or replacement vehicles should be suitable for both driving and the health and safety of the public;

- all vehicles should be regularly serviced and maintained – particularly important when pool cars are used, because pool car users often assume another user is checking on maintenance and the MOT;

- the safety critical systems (brakes, steering and tyres) need to be properly maintained;

- if the vehicle is leased and serviced by the leasing company, a system should be in place to confirm that servicing is being done to a reasonable standard;

- each vehicle should have a handbook available for the driver;

- there should be a defect reporting system for drivers to use if they consider their vehicle is unsafe; and

- if privately owned vehicles are used for work, they should be insured for business use and, if necessary, have an appropriate Safety certificate test.

For lorries

- the policy should address issues associated with the loads;

- if the load is hazardous, emergency procedures (and possibly equipment) must be in place and the driver trained in those procedures;

- the load should be stacked safely in the lorry so that it cannot move during the journey;

- there must also be satisfactory arrangements for handling the load at either end of the journey;

- the driver must be aware of the height of the vehicle, both laden and empty; and

- where tachographs are carried, they should be checked regularly to make sure that drivers are not driving for long periods without a break.

For journey planning

- journey times must include statutory breaks and allow for possible delays;

- motorways are the safest roads – minor roads are suitable for cars, but are less safe and could present difficulties for larger vehicles;

- overhead restrictions, for example bridges, tunnels and other hazards such as level crossings, may present dangers for long and/or high vehicles;

- sufficient consideration will need to be given to adverse weather conditions; and

- no alcohol or recreational drugs must be consumed during the day of the journey until the journey is completed. Only minimal amounts of alcohol should be consumed on the day before a journey is to be made.

Musculoskeletal Hazards and Risk Control

LEARNING OUTCOMES

Explain work processes and practices that may give rise to work-related upper limb disorders and appropriate control measures

Explain the hazards and control measures which should be considered when assessing risks from manual handling activities

Explain the hazards, precautions and procedures to reduce the risk in the use of lifting and moving equipment, with specific reference to manually operated load moving equipment

Explain the hazards and the precautions and procedures to reduce the risk in the use of lifting and moving equipment, with specific reference to mechanically operated load moving equipment

🔑 KEY REVISION POINTS

Musculoskeletal health issues and ill-health prevention strategies ☐

The types of injuries resulting from manual handling operations ☐

The development of manual handling risk assessments and effective control measures ☐

Safety in the use of lifting equipment and an understanding of the different types of manual and mechanical handling and lifting equipment ☐

Hazards and controls associated with fork lift trucks, hoists and cranes ☐

The requirements for the installation, examination, testing and operation of lifting equipment ☐

WORK-RELATED UPPER LIMB DISORDERS

Ergonomics – interaction between the worker, his work and his environment. It involves knowledge of his physical and mental capabilities in addition to his understanding of the job.

Musculoskeletal disorders (MSDs) affect muscles, joints, tendons, ligaments and nerves. MSDs usually affect the back, neck, shoulders and upper limbs (termed work-related upper limb disorders – WRULDs).

ILL-HEALTH EFFECTS DUE TO POOR ERGONOMICS

Work related upper limb disorders (WRULDs):

- Tenosynovitis
- Repetitive strain injury (RSI)
- Carpal tunnel syndrome
- Frozen shoulder.

Often caused by repetitive operations – keyboard operations, brick laying, assembly of small components.

DISPLAY SCREEN EQUIPMENT (DSE)

Ill health hazards from DSE work:

■ Musculoskeletal problems

■ Visual problems

■ Psychological problems.

The DSE risk assessment should consider the following factors:

■ the height and adjustability of the monitor

■ the adjustability of the keyboard, the suitability of the mouse and the provision of wrist support

■ the stability and adjustability of the DSE user's chair

■ the provision of ample foot room and suitable foot support

■ the effect of any lighting and window glare at the workstation

■ the storage of materials around the workstation

■ the safety of trailing cables, plugs and sockets

■ environmental issues – noise, heating, humidity and draughts.

MANUAL HANDLING HAZARDS, RISKS AND CONTROL MEASURES

Manual handling can involve any load movement by human effort only (lifting, pushing, pulling, carrying or supporting).

HIERARCHY OF MEASURES FOR MANUAL HANDLING OPERATIONS

1 avoid manual handling, if possible;
2 mechanise or automate the lifting process, if possible;
3 if unavoidable, make improvements to the task, load and working environment and undertake a risk assessment.

Manual handling hazards:

■ lifting a load which is too heavy and/or cumbersome

■ poor posture and/or poor lifting technique

■ dropping the load on the foot

■ lifting sharp-edged or hot loads.

Injuries caused by manual handling include:

■ muscular sprains or strains

■ back injuries

■ trapped nerve

■ hernias

■ cuts, bruises and abrasions

■ fractures

■ work-related upper limb disorders (WRULDs)

■ rheumatism.

Manual handling risk assessments require the use of mechanical aids. If this is not possible, then the assessment of:

■ the task

■ the individual

■ the load

■ the working environment.

Manual handling training includes:

■ the types of injury;

■ the manual handling assessment findings;

■ potentially hazardous manual handling operations;

- the correct use of manually operated load moving equipment, such as sack and pallet trucks;

- the correct use of mechanical aids;

- the correct use of personal protective equipment;

- good housekeeping issues;

- the factors that can affect an individual; and

- a good lifting technique, such as:

  - Check that suitable clothing (including gloves, if required) and footwear are being worn. Make an approximate assessment of the weight of the load and decide whether to lift it alone. Check to see whether one side of the load is heavier than the other and ensure that the heaviest side of the load is closest to the body.

  - Place the feet apart and adopt a good posture by bending the knees and keeping the back straight.

  - Get a firm grip and hold the load as close as possible to the body.

  - Ensuring that the back remains straight, lift the load to knee level and then to waist level without jerking.

  - After ensuring that full visibility is available, move forward without twisting the trunk.

  - Set the load down either at waist level or by lowering it first to knee level and then to the floor.

  - Manoeuvre the load to its final position after it has been set down.

MANUALLY OPERATED LOAD HANDLING EQUIPMENT

TYPES OF MANUALLY OPERATED LOAD HANDLING EQUIPMENT

- simple tools

- wheelbarrows

- trucks and trolleys

- roller tracks and chutes

- pallet trucks

- conveyors

- various types of hoists that can be used to lift people as well as other loads.

HAZARDS ASSOCIATED WITH MANUALLY OPERATED LOAD HANDLING EQUIPMENT

- incorrect use, such as overloading or attempting to carry unstable loads

- manual handling hazards due to pushing or pulling a truck or trolley

- lack of maintenance

- injuries to pedestrians using the same walkway.

PRECAUTIONS WITH THE USE OF MANUALLY OPERATED LOAD HANDLING EQUIPMENT

- when handling aids are being selected, the subsequent user should be consulted whenever possible;

- moving and handling tasks are often made easier by good design;

- a safe system of work should be in place;

- all operators of mechanical handling equipment must be properly trained in its use and supervised while they are using it;

- sufficient room available to manoeuvre the equipment easily;

- adequate visibility and lighting available;

- the floor is in a stable condition;

- regular safety checks to identify any faults with the equipment;

- a regular maintenance schedule for the equipment;

- the proposed use will be within the safe working load of the equipment;

- the lifting equipment is safety kite-marked; and

- advice is sought from the suppliers/hirers on its suitability for the proposed task and any maintenance requirements.

MECHANICALLY OPERATED LOAD HANDLING EQUIPMENT

POSITIONING AND INSTALLATION OF LIFTING EQUIPMENT

Risks during lifting operations should be reduced by:

- avoiding the equipment or its load striking a person

- avoiding a load drifting, falling freely or being released unintentionally

- avoiding the need to lift loads over people

- stopping safely in the event of a power failure

- where possible, enclosing the path of the load with suitable and substantial interlocked gates.

THE ORGANISATION OF LIFTING OPERATIONS

Every lifting operation, which is lifting or lowering a load, shall be:

- properly planned by a competent person

- appropriately supervised

- performed in a safe manner.

SUMMARY OF THE REQUIREMENTS FOR LIFTING OPERATIONS

There are four general requirements for all lifting operations:

- strong, stable and suitable lifting equipment;

- the equipment should be positioned and installed correctly;

- the equipment should be visibly marked with the safe working load; and

- lifting operations must be planned, supervised and performed in a safe manner by competent people.

TYPES OF MECHANICAL HANDLING AND LIFTING EQUIPMENT

CONVEYOR

There are three types of conveyor – belt, roller and screw.

Typical hazards include:

- in-running nips

- entanglement

- loads falling from conveyor

- impact against overhead systems

- contact hazards – sharp edges

- manual handling hazards

- noise and vibration hazards.

ELEVATOR

An elevator transports goods between floors. The most common hazard is caused by loads falling from the elevator. Also there are manual handling hazards at either end of the elevator.

FORK LIFT TRUCKS

Hazards include:

- overturning

- overloading

- collisions with pedestrians or other vehicles or structures

- the silent operation of electrically powered fork lift trucks makes pedestrians unaware of their presence

- uneven road surface

- overhead obstructions

- loss of load

- failure of the hydraulic system

- inadequate maintenance

- lack of driver training

- use as a work platform and/or carrying passengers

- speeding

- poor vision around load

- dangerous stacking technique, particularly on warehouse racking

- fire – either when battery charging or refuelling.

There are also the following physical hazards:

- noise

- exhaust fumes

- vibrations

- manual handling

- ergonomic – musculoskeletal injuries due to soft tyres, expansion joints.

Driver daily checks:

- tyres and tyre pressures

- brakes

- reversing horn and light

- mirrors

- secure and properly adjusted seat

- correct fluid levels
- correct working of all lifting and tilting systems.

CRANES

Principles of safe operation:

- brief inspection prior to each use (including lifting tackle);
- loads not left suspended when crane not in use;
- prior to lift ensure that nobody can be struck by load or crane;
- never carry loads over people;
- good visibility and communications;
- only lift loads vertically – do not drag load;
- travel with load as close to the ground as possible; and
- switch off power to crane when unattended.

Mobile jib cranes:

- plan each lift (maximum load and radius of operation);
- identify overhead obstructions and hazards;
- assess load-bearing capacity of ground; and
- if fitted, outriggers should be used.

Reasons for crane failure:

- overloading
- poor slinging of load
- overturning
- collision with another structure or overhead power lines
- foundation failure
- structural failure of the crane

- operator error

- lack of maintenance and/or regular inspections.

During lifting operations, ensure that:

- driver has good visibility;

- there are no pedestrians below load and barriers are in place; and

- audible warning given prior to lifting operation.

LIFTS, HOISTS AND ITEMS OF LIFTING TACKLE

A **lift or hoist** incorporates a platform or cage. Its movement is restricted by guides and it may carry passengers and/or goods alone.

They are required to:

- be of sound mechanical construction;

- have interlocking doors or gates that must be completely closed before the lift or hoist moves;

- be fitted, if carrying passengers, with an automatic braking system to prevent overrunning and a safety device to support the lift in the event of suspension rope failure;

- be rigorously maintained by competent persons; and

- be protective of others during maintenance operations from falling down the lift shaft and other hazards.

Other **items of lifting tackle** include chain slings and hooks, wire and fibre rope slings, eyebolts and shackles.

Important points include:

- loads must be properly secured and balanced in slings;

- lifting hooks should be checked for wear and hook distortion;

- shackles and eyebolts must be correctly tightened;

- slings must be checked for any damage before use and only used by competent people;

- training and instruction should be given in the use of lifting tackle;

- regular inspections of tackle should be made in addition to the mandatory thorough examinations; and

- lifting items should be carefully stored between uses.

The equipment should be inspected at suitable intervals between thorough examinations. The frequency and the extent of the inspection are determined by the level of risk presented by the lifting equipment. A report or record should be made of the inspection which should be kept until the next inspection. Unless stated otherwise, lifts and hoists should be inspected every week.

ILO RECOMMENDATIONS ON THE USE OF LIFTING EQUIPMENT

INSTALLATION

Fixed lifting appliances should be installed:

- by competent persons;

- so that they cannot be displaced by the load, vibration or other influences;

- so that the operator is not exposed to danger from loads, ropes or drums; and

- so that the operator can either see over the zone of operations or communicate with all loading and unloading points by telephone, signals or other adequate means.

A clearance of at least 60 cm or more should be provided between moving parts or loads of lifting appliances and:

- fixed objects in the surrounding environment such as walls and posts; or

- electrical conductors (more for high voltage transmission lines).

OTHER POINTS

1 The strength and stability of lifting appliances should take into account the effect of any wind forces to which they may be exposed.
2 No structural alterations or repairs should be made to any part of a lifting appliance which may affect the safety of the appliance without the permission and supervision of the competent person.

EXAMINATIONS AND TESTS

An **inspection** is performed by a competent person appointed by the employer and is used to identify whether the equipment can be operated, adjusted and maintained safely. Lifting appliances and items of lifting gear, as prescribed by national laws or regulations, should be examined and tested by a competent person:

a before being taken into use for the first time;
b after erection on a site;
c subsequently at intervals prescribed by national laws and regulations; and
d after any substantial alteration or repair.

The manner in which the examinations and tests are to be carried out by the competent person and the test loads to be applied for different types of lifting appliances and lifting gear should be in accordance with national laws and regulations. The results of the examinations and tests on lifting appliances and lifting gear should be recorded in prescribed forms and in conformity with national laws.

OPERATION

No lifting appliance should be operated by a worker who:

■ is below 18 years of age;

■ is not medically fit; or

■ has not received appropriate training in accordance with national laws and regulations or is not properly qualified.

Work Equipment Hazards and Risk Control

2.4

LEARNING OUTCOMES

Outline the general requirements for work equipment ■

Explain the hazards and controls for hand-held tools ■

Describe the main mechanical and non-mechanical hazards
of machinery ■

Explain the main control measures for reducing risks from
machinery hazards ■

 KEY REVISION POINTS

The definition of work equipment ☐

The safe use and maintenance of work equipment ☐

Examination and testing of pressure systems ☐

The hazards and controls related to hand-held tools ☐

Examples of mechanical and non-mechanical machinery hazards ☐

Hierarchy of measures to reduce the risks of injury from work
equipment ☐

Practical safeguards and their application to a range of machines ☐

The construction requirements of guards ☐

GENERAL REQUIREMENTS OF WORK EQUIPMENT

USE OF WORK EQUIPMENT

Definition of work equipment:

Any equipment used by an employee at work, such as:

- hand tools – hammers, knives, spanners

- machines – photocopiers, drilling machines, vehicles

- apparatus – laboratory equipment

- lifting equipment – fork lift trucks, hoists

- other equipment – ladders, vacuum cleaners

- installations – scaffolds, sound enclosures.

Uses of work equipment include:

- starting and stopping

- repair and maintenance

- modification and servicing
- cleaning
- transporting.

MANAGEMENT DUTIES

Work equipment must be:

- suitable for its purpose of use and only used for specified operations;
- suitable in terms of initial integrity, meeting basic safety standards and place of use;
- selected with the health and safety of the user in mind;
- maintained in an efficient working order and in good repair;
- inspected at regular intervals;
- restricted in use to designated persons when it has specific hazards;
- used together with appropriate information, instruction and training; and
- if appropriate, in conformity with EU requirements.

Other duties include:

- the guarding of dangerous parts of work equipment;
- the provision of stop and emergency stop controls;
- ensuring that the equipment is stable;
- the lighting around the equipment is suitable and sufficient; and
- suitable warning markings or devices, such as flashing lights, are fitted.

HIERARCHY OF RISK CONTROL FOR WORK EQUIPMENT

1. eliminate the risks
2. control risks by physical methods (such as guarding)
3. software measures (such as safe systems of work).

INFORMATION, INSTRUCTION AND TRAINING

This should include:

■ the significant health and safety issues with the equipment

■ any limitations or problems with the use of the equipment

■ safe methods to deal with these problems

■ details of any residual risks associated with the equipment

■ details of safe working procedures

■ the location of the manufacturer's manual or guidance.

Induction training will be required on recruitment but refresher training will also be needed when:

■ jobs are changed, particularly if the level of risk increases;

■ new technology or equipment is introduced;

■ the system of work is modified;

■ there are legislative changes; or

■ periodically as a refresher course.

TYPES OF MAINTENANCE

1 Preventative planned maintenance
2 Condition-based maintenance
3 Breakdown-based maintenance.

HAZARDS ASSOCIATED WITH MAINTENANCE

■ lack of competence and/or training

■ equipment not made safe before maintenance

■ lack of permit-to-work

■ use of incorrect tools or unsafe equipment.

INSPECTION OF WORK EQUIPMENT

- after installation or put into service for the first time;

- after assembly at a new site or in a new location; and thereafter

- at suitable intervals; and

- each time exceptional circumstances occur that could affect safety.

STATUTORY REQUIREMENTS FOR EXAMINATION OF BOILERS AND AIR RECEIVERS

- supplied with correct written information and markings;

- properly installed;

- used within its operating limits;

- written scheme for periodic examination certified by a competent person;

- examined by a competent person in accordance with the written scheme;

- a report of the periodic examination held on file and any required actions undertaken.

EQUIPMENT CONTROLS

The equipment should have:

- operating controls – easily reached, operated and with adequate markings and warning signs

- no obstructions or debris around it

- emergency controls (red emergency stop buttons)

- stability during use.

OTHER ISSUES

- the level and quality of general and local lighting

- clear and durable markings on equipment.

OPERATOR RESPONSIBILITIES

- comply with prescribed health and safety measures;

- take all reasonable steps to eliminate or control hazards or risks to themselves and to others;

- use of protective clothing, facilities and equipment provided to protect them from those hazards;

- report immediately to their supervisor any hazard or risk to their health and safety or that of other persons which they cannot properly deal with themselves;

- co-operate with the employer and other workers to permit compliance with the duties and responsibilities placed on the employer and workers by national laws and regulations; and

- keep the area around the machine clear and free of obstructions.

HAND TOOLS

HAND–HELD TOOLS

Typical hazards include:

- broken handles

- incorrect use of knives, saws and chisels

- tools that slip

- splayed spanners

- flying particles

- electrocution and/or burns.

Controls:

- suitable for purpose (e.g. insulated tools for electricians)

- inspection on a regular basis

- training for all hand tool users.

HAND-HELD POWER TOOLS

Hazards include:

- entanglement
- flying particles (eye injury)
- contact with cutting edges
- striking hidden/buried services (electrical cables, gas mains)
- manual handling injuries
- hand–arm vibration syndrome
- noise levels
- dust levels
- trailing leads
- fire and/or explosion.

Typical controls:

- read operating instructions before initial use
- maintain a clean, tidy and well-lit working area
- never use power tools near water, combustible fluids or gases
- store tools in a dry, safe and secure place
- never overload tools
- always use the correct tool and/or attachment for a heavy job
- always wear suitable work clothes and appropriate personal protective equipment
- ensure that the work piece is securely held in clamps or a vice
- maintain tools with care, keep them clean and sharp and inspect cables regularly
- always check that the power switch is turned off before connecting the power cable

■ use only tool accessories and attachments recommended by the tool manufacturer.

The use of chainsaws is very hazardous because:

■ there is no guard on the cutting chain

■ they are heavy

■ they are noisy.

MACHINERY HAZARDS

MECHANICAL MACHINERY HAZARDS:

■ crushing

■ shearing

■ cutting or severing

■ entanglement

■ drawing-in or trapping (in-running nips)

■ impact

■ stabbing or puncture

■ friction or abrasion

■ high pressure fluid ejection.

NON-MECHANICAL MACHINERY HAZARDS INCLUDE:

■ access to the machinery

■ slips, trips and falls

■ falling and moving objects from the machinery

■ obstructions and projections

■ manual handling and lifting

■ electricity (shock, burns or fire)

- fire and explosion

- noise and vibration

- high pressure – ejection from hydraulic hose leaks

- high/low temperature

- dust, fume or mist

- radiation – ionising and non-ionising.

Examples of equipment with machinery hazards:

- Photocopier

- Document shredder

- Bench-top grinding machine

- Pedestal drill

- Cylinder mower

- Strimmer/Brush-cutter

- Chainsaw

- Compactor

- Checkout conveyor system

- Cement mixer

- Bench-mounted circular saw.

CONTROL MEASURES FOR REDUCING RISKS FROM MECHANICAL HAZARDS

Practical safeguarding

Defined by the legal definition of 'practical' – if it is technically possible to guard an item of equipment then it should be guarded irrespective of cost and convenience.

Levels of protection or hierarchy of measures

1 Fixed enclosing guarding
2 Other guards or protection devices (interlocked, trip etc.)
3 Protection appliances (jigs, push-sticks etc.)
4 Provision of information, instruction, training and supervision.

TYPES OF GUARD

Guard	Example
Fixed	around a fan
Adjustable	around a drill bit on a pedestal drill
Interlocking	on the gates of a passenger lift, microwave oven
Trip device	photoelectric cell in front of a guillotine
Two-handed control	crimping machine
Hold-to-run	Tube train driver

Apply the above to the 11 pieces of equipment listed above – photocopier, document shredder, bench-top grinding machine, pedestal drill, cylinder mower, strimmer/brush-cutter, chainsaw, compactor, checkout conveyor system, cement mixer and a bench-mounted circular saw.

FACTORS IN GUARD CONSTRUCTION

■ Strength

■ Weight and size

■ Compatibility with process

■ Hygiene

■ Visibility

■ Noise attenuation

- Free flow of air (ventilation and/or cooling)
- Free of sharp edges
- Maintenance and cleaning
- Openings.

Electrical Safety

LEARNING OUTCOMES

Outline the principles, hazards and risks associated with the use of electricity in the workplace

Outline the control measures that should be taken when working with electrical systems or using electrical equipment in normal workplace conditions

KEY REVISION POINTS

The basic electrical principles and definitions ☐

The causes and treatment of electric shock ☐

Other electrical hazards and injuries and their associated controls ☐

The selection, suitability, maintenance and inspection of electrical
equipment ☐

The advantages and limitations of protective systems ☐

The hazards and controls associated with buried and overhead
electrical services ☐

The application, advantages and limitations of portable electrical
appliance testing ☐

HAZARDS AND RISKS ASSOCIATED WITH
THE USE OF ELECTRICITY AT WORK

ELECTRICAL PRINCIPLES AND DEFINITIONS

- voltage (volts V)

- current (amps A)

- resistance (ohms Ω)

- Ohm's law – $V = I \times R$

- Power $P = V \times I$ Watts

- conductors – electricity will flow easily

- insulators – very poor conductors

- earthing – connection to earth

- short circuit – direct flow to earth

- low voltage – not exceeding 600 V

- high voltage – exceeding 600 V
- mains voltage – 220/240 V

ELECTRICAL HAZARDS AND INJURIES

- electric shock
- electric burns
- electric fires and explosions
- arcing
- portable electrical equipment
- secondary hazards (falls, trips, noise and vibration).

ELECTRIC SHOCK

- The effect of an electric shock can vary from a slight tingling sensation to death.
- The severity of the shock depends on the size of the current and voltage.
- Low-voltage shocks, below 110 V, are seldom fatal.
- Electric shock can also produce burns.

Treatment of electric shock:

- Raise the alarm
- Switch off the power – if not possible, use insulating material to move victim from contact with the power supply
- Call for an ambulance
- If breathing, place victim in recovery position
- If not breathing, apply mouth-to-mouth resuscitation
- Treat any burns
- If person regains consciousness, treat for normal shock
- Remain with person until help arrives.

If high voltages are involved, inform police and electrical supply company but do not approach the victim.

ELECTRICAL FIRES AND EXPLOSIONS

Twenty-five per cent of all fires have an electrical origin.

Causes:

- short circuits
- overheating of cables and equipment
- ignition of flammable gases and vapours
- ignition of combustible substances by static electrical discharges
- electric arcing
- static electricity – lightning strikes.

PORTABLE ELECTRICAL EQUIPMENT

Twenty-five per cent of all electrical accidents are caused by portable electrical equipment.

Hazards:

- faulty cables, extension leads, plugs and sockets
- inadequate maintenance
- use in flammable or damp atmospheres
- misuse of equipment.

SOURCES OF HIGH RISKS ASSOCIATED WITH THE USE OF ELECTRICITY

- working with poorly maintained electrical equipment
- using electrical equipment in adverse or hazardous environments such as wet, flammable or explosive atmospheres
- working on mains electricity supplies

- contact with underground cables during excavation work

- contact with live overhead power lines.

CONTROL MEASURES

The main control measures require the following issues to be addressed:

- the design, construction and maintenance of electrical systems, work activities and protective equipment;

- the strength and capability of electrical equipment;

- the protection of equipment against adverse and hazardous environments;

- the insulation, protection and placing of electrical conductors;

- the earthing of conductors and other suitable precautions;

- the integrity of referenced conductors;

- the suitability of joints and connections used in electrical systems;

- means for protection from excess current;

- means for cutting off the supply and for isolation;

- the precautions to be taken for work on equipment made dead;

- working on or near live conductors;

- adequate working space, access and lighting; and

- the competence requirements for persons working on electrical equipment to prevent danger and injury.

SPECIFIC CONTROLS FOR ELECTRICAL HAZARDS

- A management system that ensures safe installation, operation and maintenance

- Training at all levels – induction, supervisory and technical

- Safe systems of work include permits to work on live electricity

- Selection of suitable equipment for the work and environment by considering the following:

    - atmosphere (flammable, damp)

    - weather conditions

    - high or low temperatures

    - dirty or corrosive conditions

    - standard of installation

    - rated operating conditions

- The use of protective systems

- Inspection and maintenance strategies.

Protective systems:

- Fuse (and circuit breaker)

- Insulation

- Isolation

- Reduced low-voltage systems

- Residual current devices

- Double insulation.

PROTECTION AGAINST CONTACT WITH LIVE OVERHEAD POWER LINES

As a general rule no vehicles, plant or equipment should be brought closer than:

- 15 m of overhead lines suspended from steel towers

- 9 m of overhead lines supported on wooden poles.

If a closer approach is required, lines should be made dead and a permit-to-work used.

If vehicles need to move beneath power lines, the roadway should be covered by goalposts covered with warning tape. Bunting should be suspended, level with the top of the goalposts and just above ground level between poles across the site along the length of the power line. Suitable warning signs should be placed on either side of the roadway on each side of the power line to warn of the overhead power line.

PROTECTION AGAINST CONTACT WITH LIVE BURIED POWER LINES

- Check for any obvious signs of underground services, for example valve covers or patching of the road surface.

- Ensure that the excavation supervisor has the necessary service plans and is competent to use them to locate underground services.

- Ensure that all excavation workers are trained in safe digging practices and emergency procedures.

- Use locators to trace any services and mark the ground accordingly. A series of trial holes should be dug by hand to confirm the position of the pipes or cables. This is particularly important in the case of plastic pipes which cannot be detected by normal locating equipment.

- In areas where underground services may be present, only hand digging should be used with insulated tools. Spades and shovels should be used rather than picks and forks, which are more likely to pierce cables.

- Assume that all cables are 'live' unless it is known otherwise.

- Hand-held power tools should not be used within 0.5 m of the marked position of an electricity cable. Collars should be fitted to the tools so that initial penetration of the surface is restricted.

- Any suspected damage to cables must be reported to the service providers and the health and safety enforcement authority.

- All exposed cables should be backfilled with fine material such as dry sand or small gravel.

- The service plans must be updated when the new cables have been laid.

INSPECTION AND MAINTENANCE STRATEGIES

The particular areas of interest for inspection and maintenance are:

- the competence of the inspection and maintenance team

- the cleanliness of insulator and conductor surfaces

- the mechanical and electrical integrity of all joints and connections

- the integrity of mechanical mechanisms, such as switches and relays

- the calibration, condition and operation of all protection equipment, such as circuit breakers, RCDs and switches

- isolation procedures

- inspection and maintenance records

- permits to work for work on live electricity.

INSPECTION STRATEGIES

Any strategy for the inspection of electrical equipment, particularly portable appliances, should involve the following considerations:

- a means of identifying the equipment to be tested

- the number and type of appliances to be tested

- the competence of those who will undertake the testing (whether in-house or bought-in)

- the legal requirements for portable appliance testing (PAT) and other electrical equipment testing and the guidance available

- organisational duties of those with responsibilities for PAT and other electrical equipment testing

- test equipment selection and re-calibration

- the development of a recording, monitoring and review system

- the development of any training requirements resulting from the test programme.

PORTABLE ELECTRICAL APPLIANCE TESTING

General points:

- user checks – at least once a week

- formal visual inspection – by a trained person

- combined testing and inspection – by a competent person using testing equipment

- the frequency of inspection and testing depends on usage rate and conditions – normally determined by a risk assessment

- records of inspection and testing required.

Advantages of portable appliance testing	Limitations of portable appliance testing
• early indication of serious faults • discovery of incorrect equipment or electrical supply • incorrect fuses found • reduction in the number of electrical accidents • monitoring misuse of portable appliances • check on selection procedures • increase awareness of hazards • introduction of a more regular maintenance regime	• too frequent testing of fixed equipment can lead to excessive costs • some unauthorised equipment never tested because of lack of records • equipment misused or overused between tests • the inclusion of trivial faults leads to long lists and the overlooking of significant faults • the level of competence of tester • the use of incorrectly calibrated test equipment

Fire Safety

2.6

LEARNING OUTCOMES

Describe the principles of fire initiation, classification and spread ☐

Outline the principles of fire risk assessment ☐

Describe the basic principles of fire prevention and the prevention of fire spread in buildings ☐

Identify the appropriate fire alarm systems and fire-fighting equipment for a simple workplace ☐

Outline the factors which should be considered when implementing a successful evacuation of a workplace in the event of a fire ☐

KEY REVISION POINTS

The fire triangle ☐

The legal framework for the control of fire hazards ☐

The basic principles of fire prevention – classification, methods of
extinction and principles of heat transmission ☐

The common causes of fire, its spread and consequences ☐

Key elements of a fire risk assessment ☐

The principles of fire detection, protection, warning and means
of escape ☐

The provision, maintenance and testing of fire-fighting
equipment ☐

Emergency planning and training ☐

FIRE INITIATION, CLASSIFICATION AND SPREAD

BASIC PRINCIPLES

The fire triangle consists of:

- Fuel – solids, liquids and gases

- Ignition source

- Oxygen – from air, oxidising materials.

CLASSES OF FIRE

- A – carboniferous solids

- B1 – liquids soluble in water

- B2 – liquids insoluble in water

- C – gases

- D – metals
- F – high temperature cooking oils or fats.

Electrical fires do not constitute a separate fire class.

SOURCES OF IGNITION

- Naked flames
- External sparks
- Internal sparking
- Hot surfaces
- Static electricity.

METHODS OF EXTINCTION

- Cooling – reducing ignition temperature
- Smothering – limiting oxygen available
- Starving – limiting fuel supply
- Chemical reaction – interrupting the chain of combustion.

PRINCIPLES OF HEAT TRANSMISSION AND FIRE SPREAD

- Convection – through air movements (floor to floor)
- Conduction – through conducting materials (girders)
- Radiation – emission of heat waves (bar heaters)
- Direct burning – contact with flames (rubbish catching fire).

CAUSES OF FIRE

- rubbish (particularly if burnt)
- electrical short circuits

- portable heaters
- friction causing sparks (hand grinders)
- oil heating plant
- cooking appliances
- arson (often by children)
- hot surfaces (welding)
- cigarette smoking.

Fire spreads via:

- gases/sprays
- foam plastics
- packaging
- previous spillage
- flammable dusts.

Fire spreads due to:

- delayed discovery
- alarm not raised
- fire service not informed
- lack of barriers
- fire procedures not known
- fire extinguishers unavailable, not serviced or not used
- air flow – particularly if fire doors not closed
- lack of or poor emergency lighting
- flammable materials
- flammable dusts.

FIRE RISK ASSESSMENT

STAGES OF THE FIRE RISK ASSESSMENT

1 Identify the fire hazards – identify sources of heat/ignition, unsafe acts (smoking) and unsafe conditions (storage of combustible materials);
2 Identify all locations and persons at risk;
3 Reduce the risks by the introduction of controls;
4 Record the findings; and
5 Monitor and review.

Hazards	Other factors
• sources of fuel • oxygen depletion • flames and heat • smoke • gaseous combustion products • structural failure of the building	• time to detect and warn • safe egress • adequacy of fire signs • provision of fire-fighting equipment • fire training – fire drills, appliance use and emergency teams • maintenance of fire equipment • consultation with employees

TEMPORARY WORKPLACES

Construction sites, temporary classrooms and buildings, exhibitions, festivals and fetes.

The requirements depend on:

■ the size of the premises;

■ the number of persons working or visiting the site at any one time;

■ the degree of flammability of the materials used to construct the workplace or being used in the workplace;

■ height above or below the ground floor and distance to a place of safety; and

■ the remoteness of the workplace relative to water supplies and/or fire and rescue services.

FIRE PREVENTION AND PREVENTION OF FIRE SPREAD

CONTROL MEASURES

- elimination or reduction in the use and storage of flammable materials
- electrical safety
- control of ignition sources
- effective systems of work
- safe refuelling procedures for vehicles
- safe heating and cooking equipment and procedures
- good housekeeping
- storage facilities for small quantities (50 litres) of highly flammable or flammable liquids.

FIRE ALARM SYSTEM AND FIRE-FIGHTING EQUIPMENT

FIRE EXTINGUISHERS

Fire extinguishers are normally coloured red over much % of their surface.

Colour (5%)	Contents	Class/type of fire
Red	Water	A
Cream	AFFF	A (possibly B)
Cream	Foam	B and C
Black	Carbon dioxide	B and electrical
Blue	Dry powder	B, C and electrical

OTHER FIRE-FIGHTING EQUIPMENT

- fire blankets
- hose reels
- sprinklers.

EMERGENCY LIGHTING

■ Regular testing required.

SITING, MAINTENANCE AND TESTING OF FIRE-FIGHTING EQUIPMENT

■ sited in areas of significant risk (at least one extinguisher per 200 m^2 of floor space with a minimum of one per floor)

■ arrangements for hot working

■ annual maintenance check of fire extinguishers

■ weekly check of fire alarms and emergency lighting.

Fire training in the use of the equipment for all staff is essential.

FIRE DETECTION AND WARNING

■ fire notices

■ alarms – including smoke and heat detectors

■ audibility throughout premises

■ regularly tested

■ records of all tests

■ vigilance of workers following training.

EVACUATION OF THE WORKPLACE

MEANS OF ESCAPE

■ Doors – clear of obstructions and opening in direction of escape

■ Escape routes (stairways and passageways) directly to a place of safety

■ Lifts should not be used

■ Lighting, including emergency lighting

■ Directional and other signs

- Escape times
- Assembly points
- Building plans to include record of emergency escape plans.

STRUCTURAL DESIGN MEASURES

- Fire loading of the building
- Surface spread of fire
- Fire resistance of structural elements
- Insulating materials
- Fire compartmentation.

EMERGENCY PLANNING AND TRAINING

- Evacuation procedures – fire notices
- Appointment of fire marshals
- Fire drills
- Roll call
- Provision for the infirm and disabled.

FIRE PROCEDURES FOR PEOPLE WITH A DISABILITY

Emergency planning needs to consider the special requirements of people with physical or mental disabilities. Such issues include:

- the identification of those who may need special help to evacuate the building;
- the allocation of responsibility to specific staff to identify and assist people with a disability during emergencies;
- plan suitable escape routes;

■ have a system in place that enables a person with a disability to summon help in emergencies; and

■ include the arrangements for disabled persons in all staff fire training programmes.

GENERAL POINTS

■ Emergency procedures required for fire, explosions, flooding or structural collapse.

■ A hot work permit should include the need for a fire extinguisher and checks for signs of fire on completion of the work.

■ A hot work permit would be needed when cutting steel with an angle grinder and soldering pipe work in a central heating system.

■ Frost around the valve on a Liquid Petroleum Gas cylinder indicates that the valve is leaking.

■ Cylinders containing Liquid Petroleum Gas should be stored in a locked external compound at least 3 m from any oxygen cylinders.

■ The minimum distance that oxygen cylinders should be stored away from propane, butane and other inflammable gases is 3 m.

■ Butane and propane are heavier than air and therefore represent a ground level fire hazard.

Chemical and Biological Health Hazards and Risk Control

2.7

LEARNING OUTCOMES

Outline the forms of, the classification of, and the health risks from exposure to, hazardous substances ■

Explain the factors to be considered when undertaking an assessment of the health risks from substances commonly encountered in the workplace ■

Describe the use and limitations of Workplace Exposure Limits, including the purpose of long-term and short-term exposure limits ■

Outline control measures that should be used to reduce the risk of ill health from exposure to hazardous substances ■

Outline the hazards, risks and controls associated with specific agents ■

Outline the basic requirements related to the safe handling and storage of waste ■

KEY REVISION POINTS

The forms and classification of hazardous chemical and biological agents ☐

The functions of various human body organs and their vulnerability to hazardous substances ☐

The difference between acute and chronic effects ☐

The assessment of health risks and recommended control measures ☐

The definitions and limitations of workplace exposure limits ☐

The use of engineering controls (LEV) and personal protective equipment ☐

The health risks and controls associated with asbestos and other specific agents ☐

Health surveillance ☐

Emergency and maintenance controls ☐

The safe handling and storage of waste ☐

FORMS AND CLASSIFICATION OF, AND HEALTH RISKS FROM, HAZARDOUS SUBSTANCES

AGENTS

Chemical agents	Biological agents

• Dusts – including respirable dust (fine dust that remains in the lungs)
  • Gases
• Vapours – substances very close to their boiling points (e.g. solvents)
  • Liquids
 	
• Mists – similar to vapours but closer to the liquid phase (e.g. paint sprays)
• Fume – condensed metallic particles (e.g. welding fume)
	• Fungi
	• Moulds
	• Bacteria
	• Viruses

CLASSIFICATION OF HAZARDOUS SUBSTANCES

▨ Irritant – repeated contact with skin can cause a sensitised or allergic inflammation

▨ Harmful – involves limited health risks

▨ Corrosive – may destroy living tissue

▨ Toxic – produces serious health risks

▨ Infectious – contains micro-organisms known to cause diseases

▨ Carcinogenic – may induce cancer

▨ Mutagenic – may induce hereditary genetic defects.

Acute effects are of short duration, normally reversible and appear shortly after exposure to a hazardous substance. (e.g. asthma attacks).

Chronic effects develop over a period of time after repeated exposure to a hazardous substance and are often irreversible.

ASSESSMENT OF HEALTH RISKS

CATEGORIES OF HEALTH RISKS

- Chemical – exhaust fumes, paint solvents, asbestos

- Biological – Legionella, other pathogens, hepatitis

- Physical – noise vibration, radiation

- Psychological – stress, violence

- (Ergonomic – musculoskeletal disorders).

TYPES OF HEALTH RISK

- skin contact with irritant substances, leading to dermatitis;

- inhalation of respiratory sensitizers, triggering immune responses such as asthma;

- badly designed workstations requiring awkward body postures or repetitive movements, resulting in upper limb disorders, repetitive strain injury and other musculoskeletal conditions;

- noise levels which are too high, causing deafness and conditions such as tinnitus;

- too much vibration, for example from hand-held tools leading to hand–arm vibration syndrome and circulatory problems;

- exposure to ionising and non-ionising radiation, including ultraviolet in the sun's rays, causing burns, sickness and skin cancer;

- infections ranging from minor sickness to life-threatening conditions, caused by inhaling or being contaminated with microbiological organisms; and

- stress causing mental and physical disorders.

ROUTES OF ENTRY INTO THE BODY

- Inhalation via the lungs

- Absorption via the skin

- Ingestion via the stomach

- Injection through the skin.

MAJOR HUMAN BODY SYSTEMS

There are five major functional systems within the body. These are shown together with typical diseases and related hazardous substances.

Body system	Illness or disease	Hazardous substance
Respiratory	bronchitis, asthma, fibrosis, cancer	dusts, asbestos
Nervous	anxiety, epilepsy, loss of consciousness	solvents, lead
Cardiovascular	headaches, loss of consciousness	benzene, carbon monoxide
Urinary	cirrhosis, cancer	chlorinated hydrocarbons, mercury
Skin	irritant contact dermatitis and allergic contact dermatitis	detergents, turpentine, epoxy resin

FACTORS TO BE TAKEN INTO ACCOUNT WHEN ASSESSING HEALTH RISKS

- Suitable and sufficient health risk assessment

- Adequate control of exposure of employees

- Proper use of control measures provided

- Maintain control measures

- Monitor employees exposed to hazardous substances

- Health surveillance

- Information, instruction and training.

HEALTH RISK ASSESSMENT

■ Identify hazardous substances;

■ Gather information;

■ Evaluate the risks;

■ Decide on control measures;

■ Record the assessment; and

■ Review the assessment.

SOURCES OF INFORMATION

■ Product labels

■ Safety data sheets

■ EU list of Indicative Limit Values

■ HSE list of Workplace Exposure Limits (UK)

■ ACGIH list of Threshold Limit Values (USA)

■ Trade association publications

■ Industrial codes of practice

■ Specialist reference manuals.

SURVEY TECHNIQUES FOR HEALTH RISKS

include:

■ Stain tube detector. The advantages of this technique are that it is quick, relatively simple to use and inexpensive. The disadvantages are:

  ● it cannot be used to measure concentrations of dust or fume;

  ● the accuracy of the reading is approximately $\pm 25\%$;

  ● it will yield false readings if other contaminants present react with the crystals;

- it can only give an instantaneous reading, not an average reading over the working period (TWA); and

- the tubes are very fragile with a limited shelf life.

■ Passive sampling, including the static dust sampler

■ Sampling pumps and heads

■ Direct reading instruments (MIRA)

■ Vane anemometers and hygrometers

■ Smoke tube and dust observation lamp.

WORKPLACE EXPOSURE LIMITS (WELS)

The WEL is related to the concentration of airborne hazardous substances that people breathe over a specified period of time.

The most widely used limits, called threshold limit values (TLVs), are those issued in the USA.

For airborne exposures, there are four types of limits in common use:

■ the time-weighted average (TWA) exposure limit – the maximum average concentration of a chemical in air for a normal 8-hour working day and 40-hour week;

■ the short-term exposure limit (STEL) – the maximum average concentration to which workers can be exposed for a short period (usually 15 minutes);

■ the ceiling value – the concentration that should not be exceeded at any time;

■ maximum permissible concentrations or threshold level values (TLVs).

In the **United Kingdom**, the HSE has assigned WELs to a large number of hazardous substances and publishes any updates in a publication called 'Occupational Exposure Limits' EH40.

Workplace exposure limits (WELs) must not be exceeded.

Two categories – similar to previous maximum exposure limit (MEL) and occupational exposure standard (OES):

| Category 1 | carcinogenic and mutagenic substances – exposure must be reduced as low as is reasonably practicable below the WEL |
| Category 2 | all other hazardous substances – exposure controlled by principles of good practice |

There are two time-weighted averages (TWA):

- the long-term exposure limit (LTEL) – over 8 hours
- the short-term exposure limit (STEL) – 15-minute reference period.

LIMITATIONS OF EXPOSURE LIMITS

- They are specifically quoted for an 8-hour period (with an additional STEL for many hazardous substances). Adjustments must be made when exposure occurs over a continuous period longer than 8 hours.

- They can only be used for exposure in a workplace and not to evaluate or control non-occupational exposure (e.g. to evaluate exposure levels in a neighbourhood close to the workplace, such as a playground).

- WELs are only approved where the atmospheric pressure varies from 900 to 1100 millibars. This could exclude their use in mining and tunnelling operations.

- They should also not be used when there is a rapid build-up of a hazardous substance due to a serious accident or other emergency.

- The fact that a substance has not been allocated a WEL does not mean that it is safe.

CONTROL MEASURES

The general principles of prevention are:

1 the avoidance of risks;
2 the evaluation of unavoidable risks;

3   controlling hazards at source; and
4   replacing the dangerous by less or non-dangerous alternatives;
5   adapting work to the individual;
6   adapting to technical progress;
7   developing a coherent prevention policy;
8   giving priority to collective over individual protective measures; and
9   giving appropriate instructions to employees.

PRINCIPLES OF GOOD PRACTICE FOR THE CONTROL OF EXPOSURE TO SUBSTANCES HAZARDOUS TO HEALTH

1   Design and operate processes and activities to minimise emission, release and spread of substances hazardous to health.
2   Take into account all relevant routes of exposure – inhalation, skin absorption and ingestion – when developing control measures.
3   Control exposure by measures that are proportionate to the health risk.
4   Choose the most effective and reliable control options which minimise the escape and spread of substances hazardous to health.
5   Where adequate control of exposure cannot be achieved by other means, provide, in combination with other control measures, suitable PPE.
6   Check and review regularly all elements of control measures for their continuing effectiveness.
7   Inform and train all employees on the hazards and risks from the substances with which they work and the use of control measures developed to minimise the risks.
8   Ensure that the introduction of control measures does not increase the overall risk to health and safety.

HIERARCHY OF CONTROL MEASURES

1   eliminate the risks by either designing them out or changing the process;
2   substitution of hazardous substances with others which are less hazardous;
3   isolation, using enclosures, barriers or worker segregation;
4   engineering controls such as guarding, the provision of local exhaust ventilation systems, the use of reduced voltage systems or residual current devices;

5 management controls such as safe systems of work, training, job rotation and supervision; and

6 finally, as a last resort, the provision of personal protective equipment such as ear defenders or respiratory protective equipment.

Hence, measures for preventing or controlling exposure to hazardous substances include one or a combination of the following:

■ elimination of the substance

■ substitution of the substance (or reduction in the quantity used)

■ total or partial enclosure of the process

■ local exhaust ventilation

■ dilute or general ventilation

■ reduction of the number of employees exposed to a strict minimum

■ reduced time exposure by task rotation and the provision of adequate breaks

■ good housekeeping

■ training and information on the risks involved

■ effective supervision to ensure that the control measures are being followed

■ personal protective equipment (such as clothing, gloves and masks)

■ welfare (including first aid)

■ medical records

■ health surveillance.

PREVENTATIVE CONTROL MEASURES

ENGINEERING CONTROLS

These include:

■ segregation of people from the process (fume cupboard)

■ modification of the process to reduce human contact with the hazardous substances

- local exhaust ventilation comprising:

    - a collection hood and intake

    - ventilation ducting

    - a filter or other air cleaning device

    - a fan

    - an exhaust duct

- dilution or general ventilation.

SUPERVISORY OR PEOPLE CONTROLS

Additional supervisory controls when hazardous substances are involved are:

- reduced time exposure – thus ensuring that workers have breaks in their exposure periods;

- reduced number of workers exposed – only persons essential to the process should be allowed in the vicinity of the hazardous substance. Walkways and other traffic routes should avoid any area where hazardous substances are in use;

- eating, drinking and smoking must be prohibited in areas where hazardous substances are in use;

- any special rules, such as the use of personal protective equipment, must be strictly enforced.

PERSONAL PROTECTIVE EQUIPMENT

Personal protective equipment (PPE) is the control measure of **last resort**. It is covered by the Personal Protective Equipment at Work Regulations, which require:

- PPE to be suitable for wearer and task;

- PPE to be compatible with any other personal protective equipment;

- a risk assessment to be undertaken to determine best PPE;

- a suitable PPE maintenance programme;

- suitable storage arrangements for the PPE;

- information, instruction and training for the user of the PPE;

- supervision of the use of PPE; and

- a system for reporting defects.

Notes:

1 The Regulations do not cover Respiratory Protective Equipment (RPE).

2 The risk assessment and method statement for the particular activity should include information on personal protective equipment.

Types of personal protective equipment	Types of respiratory protective equipment
• hand and skin protection	• filtering half mask
• eye protection	• half mask respirator
• protective clothing – boots, hard hats, aprons	• full face mask respirator
	• powered respirator

HEALTH SURVEILLANCE

Health surveillance is not the same as health monitoring. Types of health surveillance include:

- checking individuals on a regular basis;

- checking individuals as a result of statistical evidence of a potential problem; and

- when a substance listed in Schedule 6 of COSHH is being used. In these cases, health surveillance should take place at intervals not exceeding 12 months and records should be kept for 40 years.

MAINTENANCE AND EMERGENCY CONTROLS

Maintenance involves:

- the cleaning, testing and, possibly, the dismantling of equipment;

- the changing of filters in extraction plant or entering confined spaces;

- the handling of hazardous substances and the safe disposal of waste material;

- a permit-to-work procedure may also need to be in place, since the control equipment will be inoperative during the maintenance operations;

- the keeping of records of maintenance for at least 5 years.

The following points should be considered when **emergency procedures** are being developed:

- the possible results of a loss of control (e.g. lack of ventilation);

- dealing with spillages and leakages (availability of effective absorbent materials);

- raising the alarm for more serious emergencies;

- evacuation procedures, including the alerting of neighbours;

- fire-fighting procedures and organisation;

- availability of respiratory protection equipment; and

- information and training.

SPECIFIC AGENTS

ASBESTOS

A fine, sharp and hard fibrous dust of respirable dust size which can become lodged in the lungs, causing damage to the lining of the lungs over a period of many years. Asbestos is a possible chronic health risk. It can lead to one of the following diseases:

- asbestosis or fibrosis (scarring) of the lungs

- lung cancer

- mesothelioma – cancer of the lining of the lung.

Duty to manage asbestos

The management of asbestos in the workplace requires:

■ the determination of the location and condition of materials likely to contain asbestos (ACMs)

■ the presumption that materials contain asbestos unless there is strong counter-evidence

■ up-to-date records of the location and condition of the ACMs or presumed ACMs

■ an assessment of the risk of anyone being exposed to fibres from ACMs

■ a plan to manage the risks from ACMs

■ the necessary steps to action the plan

■ the periodic review and monitoring of the plan

■ provision of information and asbestos awareness training to anyone who is liable to work on ACMS or otherwise disturb them.

Types of asbestos survey

■ the presumptive survey – the location and assessment of condition of ACMs

■ the sampling survey – a number of samples are sent away for analysis

■ the full access and sampling survey – most invasive and detailed.

Stages in the control of asbestos

■ **Identification** of the presence of asbestos

■ **Assessment** is an evaluation as to whether the location or the condition could lead to the asbestos being disturbed

■ **Removal** must only be done by a licensed contractor

■ **Control measures** during the removal of asbestos to ensure adequate protection of workers and others

- **Medical surveillance** in the form of a regular medical examination should be given to any employee who has been exposed to asbestos at levels above the action level

- **Awareness training** is given to those employees:

  - who are or who are liable to be exposed to asbestos or who supervise such employees

  - who carry out work in connection with the employer's duties under these Regulations, so that they can carry out that work effectively

- **Disposal of asbestos waste** to an authorised asbestos waste site.

If asbestos is discovered during the performance of a contract, work should cease immediately and the employer should be informed.

HEALTH RISKS AND CONTROLS ASSOCIATED WITH OTHER SPECIFIC AGENTS

Each agent is followed by a brief description, two possible illnesses and possible controls.

Hepatitis and many blood-borne viruses can cause nausea, jaundice and death. Possible control measures:

- prohibit eating, drinking and smoking in working areas where there is a risk of contamination;

- prevent puncture wounds, cuts and abrasions, especially in the presence of blood and body fluids;

- when possible avoid use of, or exposure to, sharps such as needles, glass or metal, or if unavoidable take care in handling and disposal;

- consider the use of devices incorporating safety features, such as safer needle devices and blunt-ended scissors;

- cover all breaks in exposed skin by using waterproof dressings and suitable gloves;

- protect the eyes and mouth by using a visor/goggles/safety spectacles and a mask, where splashing is possible;

- avoid contamination by using water-resistant protective clothing;

- wear rubber boots or plastic disposable overshoes when the floor or ground is likely to be contaminated;

- use good basic hygiene practices, such as hand washing;

- control contamination of surfaces by containment and using appropriate decontamination procedures; and

- dispose of contaminated waste safely.

HIV is an example of a blood-borne virus.

Organic solvents are sensitisers and irritants that can cause dermatitis and liver failure. The minimum PPE requirements are impermeable overalls, apron, footwear, long gloves and gauntlet and chemically resistant goggles or visor. Respiratory protective equipment is also required if it cannot be demonstrated that exposure is below the appropriate workplace exposure limit (WEL). Organic solvents are often used with painting work.

Carbon monoxide is an odourless gas that can cause headaches and unconsciousness. Exposure may be prevented by ensuring that any work carried out in relation to gas appliances in domestic or commercial premises is undertaken by a competent Gas Safe Registered engineer.

Isocyanates are sensitisers and irritants that can cause bronchitis and extreme 'asthma attack'. They are often used in paint spraying operations. They require good ventilation and Respiratory Protective Equipment (RPE) should be worn.

Cement dust and wet cement can cause serious burns or ulcers and dermatitis. PPE in the form of gloves, overalls with long sleeves and full-length trousers and waterproof boots must be worn.

Wood dust can be hazardous, particularly from hard woods and composite boards (MDF), and can cause irritation to the eyes, nose and throat, leading to dermatitis, asthma, rhinitis and even cancer. An extraction system is essential and PPE in the form of gloves, respiratory protective equipment overalls and eye protection.

Silica dust produced in masonry work can cause fibrosis and silicosis. Prevention is best achieved by the use of good dust extraction systems and respiratory protective equipment.

Leptospira is a bacterium in rat urine that can cause anaemia and jaundice. Good, impervious protective clothing, particularly Wellington boots, and the covering of any skin wounds are essential in these situations.

Legionella are bacteria in tepid water and can cause pneumonia and death. Where a plant is at risk of the development of Legionella, the following is required:

- a suitable 'written and sufficient' risk assessment;

- the preparation and implementation of a written control scheme involving the treatment, cleaning and maintenance of the system;

- appointment of a named person with responsibility for the management of the control scheme;

- the monitoring of the system by a competent person; and

- record keeping and the review of procedures developed within the control scheme.

SAFE HANDLING AND STORAGE OF WASTE

The principal requirements for safe waste disposal are:

- to handle waste so as to prevent any unauthorised escape into the environment;

- to pass waste only to an officially authorised person; and

- to ensure that a written description accompanies all waste.

There are several different categories of waste. The important ones are as follows:

- Controlled waste, comprising household, industrial or commercial waste.

- Hazardous waste, which can only be disposed of using special arrangements (toxic, corrosive, carcinogenic or highly flammable).

REQUIREMENTS FOR WASTE SKIPS

- sufficient strength to cope with its load;

- stability while being filled;

- a reasonable uniform load distribution within the skip at all times;

- the immediate removal of any damaged skip from service and the skip inspected after repair before it is used again;

- sufficient space around the skip to work safely at all times;

- the skip should be resting on firm level ground;

- the skip should never be overloaded or overfilled; and

- there must be sufficient headroom for the safe removal of the skip when it is filled.

THE CONTROL MEASURES FOR HAZARDS ASSOCIATED WITH WASTE DISPOSAL

- All workers should wear suitable PPE, such as high-visibility jackets, gloves and suitable footwear.

- Training so that employees segregate hazardous and non-hazardous wastes on site and fully understand the risks and necessary safety precautions to be taken.

- The storage site should be protected against trespassers, fire and adverse weather conditions.

- If flammable or combustible wastes are being stored, adequate fire protection systems must be in place.

- Manual handling issues may need to be considered.

- All lifting equipment, including chains and shackles, must be subject to a periodic statutory examination.

- Separate storage required for incompatible waste streams.

- Finally, for liquid wastes, drains must be protected and bunds used to restrict spreading of the substance as a result of spills.

Physical and Psychological Health Hazards and Risk Control

2.8

LEARNING OUTCOMES

Outline the health effects associated with exposure to noise and appropriate control measures ☐

Outline the health effects associated with exposure to vibration and appropriate control measures ☐

Outline the health effects associated with ionising and non-ionising radiation and outline appropriate control measures ☐

Outline the causes and effects of stress at work and appropriate control measures ☐

 **KEY REVISION POINTS**

The health effects of exposure to noise and vibration ☐

The exposure limits to noise and vibration ☐

The control measures for excessive vibration and noise ☐

The hazards and controls associated with ionising and
non-ionising radiations ☐

The causes and prevention of workplace stress ☐

NOISE

HEALTH EFFECTS OF NOISE

Human ear:

- Ear drum
- Cochlea
- Hair cells – damage irreversible
- Auditory nerve.

Acute effects of noise:

- Temporary threshold shift
- Tinnitus
- Acute acoustic trauma.

Chronic effects of noise:

- Noise-induced hearing loss
- Permanent threshold shift
- Tinnitus.

NOISE ASSESSMENTS

Objective is to identify whether action levels have been reached.

NOISE MEASUREMENT

- Sound pressure level (SPL) – dB(A)

- Peak sound pressure

- Continuous equivalent noise level (L_{Eq}) – normally measured over an 8-hour period

- Daily personal exposure level ($L_{EP,d}$).

The ILO recommends that noise measurements should:

- quantify the level and duration of exposure of workers and compare it with exposure limits as established by the competent authority or internationally recognised standards;

- identify and characterise the sources of noise and the exposed workers;

- create a noise map for the determination of risk areas;

- assess the need both for engineering noise prevention and control and for other appropriate measures and for their effective implementation;

- evaluate the effectiveness of existing noise prevention and control measures.

The ILO further recommends that the level of noise and/or duration of exposure should not exceed the limits established by the competent authority or other internationally recognised standards. The assessment should consider the following:

- the risk of hearing impairment;

- the degree of interference with speech communications essential for safety purposes;

- the risk of nervous fatigue, with due consideration of the mental and physical workload and other non-auditory hazards or effects.

Summary of a noise assessment:

■ assess noise levels and keep records;

■ reduce the risks from noise exposure by engineering controls – only use hearing protection as a last resort;

■ provide employees with information and training; and

■ if a manufacturer of equipment, to provide information on noise levels of the equipment.

NOISE ACTION LEVELS

European Union noise action levels:

■ First action level – daily exposure level of 80 dB(A) – action advised

■ Second action level – daily exposure level of 85 dB(A) – action obligatory

■ Peak action level – 135 dB(C) and 137 dB (C)

■ Exposure limit values – 87 dB(A) and 140 dB(C).

To prevent the adverse effects of noise on workers, employers should:

■ identify the sources of noise and the tasks which give rise to exposure;

■ seek the advice of the competent authority and/or the occupational health service about exposure limits and other standards to be applied;

■ seek the advice of the supplier of processes and equipment about expected noise emission;

■ if this advice is incomplete or otherwise of doubtful value, arrange for measurements by persons competent to undertake these in accordance with current national and/or internationally recognised standards.

Noise control techniques include:

■ Reduction of noise at source

■ Change equipment or process

■ Change speed

- Improve maintenance

- Re-locate equipment

- Enclose equipment

- Screens or absorption walls

- Damping

- Lagging

- Silencers

- Isolation of workers

- Suitable warning signs.

Personal ear protection:

- Ear plugs

- Ear defenders (earmuffs).

Factors to be considered:

- Suitability (frequencies)

- Acceptability and comfort

- Durability

- Instruction in use

- Hygiene considerations

- Beards, hair and spectacles may reduce effectiveness of protection

- Maintenance and storage

- Cost.

VIBRATION

ILL HEALTH DUE TO VIBRATION

Ill health due to hand–arm vibration

- Hand–arm vibration syndrome (HAVS)
- Vibration White Finger (VWF)
- Carpal tunnel syndrome.

Ill health due to whole body vibration

- Severe back pain
- Reduced visual and manual control
- Increased heart rate and blood pressure
- Spinal damage.

EUROPEAN UNION EXPOSURE LIMIT AND ACTION VALUES FOR VIBRATION

A daily exposure limit and action values for both HAV and WBV has been introduced into the UK and the European Union. These values are as follows:

For hand–arm vibration (HAV)

a the daily exposure limit value normalised to an 8-hour reference period is 5 m/s^2

b the daily exposure action value normalised to an 8-hour reference period is 2.5 m/s^2.

For whole body vibration (WBV)

a the daily exposure limit value normalised to an 8-hour reference period is 1.15 m/s^2

b the daily exposure action value normalised to an 8-hour reference period is 0.5 m/s^2.

PREVENTATIVE AND PRECAUTIONARY MEASURES

The common measures used to control ergonomic ill-health effects are:

■ identification of repetitive actions;

■ the elimination of vibration by performing the job in a different way;

■ undertake a risk assessment;

■ ensure that the correct equipment (properly adjusted) is always used;

■ introduce job rotation or frequent breaks so that workers have a reduced time exposure to the hazard;

■ during the design of the job, ensure that poor posture is avoided;

■ issue employees with gloves and warm clothing;

■ examine ill-health reports and absence records;

■ introduce a programme of health surveillance;

■ ensure that workers are given adequate information on the hazards and develop a suitable training programme;

■ introduce a reporting system for employees to use so that concerns and any symptoms can be recorded and investigated;

■ ensure that drill bits and other tools are kept sharp;

■ ensure that a programme of preventative maintenance is introduced and include the regular inspection of items such as vibration isolation mountings; and

■ keep up to date with advice from equipment manufacturers, trade associations and health and safety sources.

ADDITIONAL INFORMATION ON WHOLE BODY VIBRATION

The reasons for back pain in drivers include:

■ poor posture while driving

■ incorrect adjustment of the driver's seat

- difficulty in reaching all relevant controls due to poor design of the controls layout

- frequent manual handling of loads

- frequent climbing up and down from a high cab.

Actions for controlling the risks from WBV need to ensure that:

- the driver's seat is correctly adjusted so that all controls can be reached easily and that the driver weight setting on the suspension seat, if available, is correctly adjusted. The seat should have a back rest with lumbar support;

- anti-fatigue mats are used if the operator has to stand for long periods;

- the speed of the vehicle is such that excessive jolting is avoided. Speeding is one of the main causes of excessive WBV;

- all vehicle controls and attached equipment are operated smoothly;

- only established site roadways are used;

- only suitable vehicles and equipment are selected to undertake the work and cope with the ground conditions;

- the site roadway system is regularly maintained;

- all vehicles are regularly maintained, with particular attention being paid to tyre condition and pressures, vehicle suspension systems and the driver's seat;

- work schedules are regularly reviewed so that long periods of exposure on a given day are avoided and drivers have regular breaks;

- prolonged exposure to WBV is avoided for at risk groups (older people, young people, people with a history of back problems and pregnant women); and

- employees are aware of the health risks from WBV, the results of the risk assessment and the ill-health reporting system. They should also be trained to drive in such a way that excessive vibration is reduced.

ILO RECOMMENDATIONS ON HEALTH SURVEILLANCE

Workers should be given a pre-employment medical examination for jobs involving hand–arm vibration to check whether Raynaud's phenomenon of non-occupational origin or hand–arm vibration syndrome (HAVS) is present from earlier employment. Where these symptoms are diagnosed, such employment should not be offered unless vibration has been satisfactorily controlled.

If a worker is exposed to hand-transmitted vibration, the occupational health professional responsible for health surveillance should:

(a) examine the worker periodically, as prescribed by national laws and regulations, for HAVS and ask the worker about any symptoms;
(b) examine the worker for symptoms of possible neurological effects of vibration, such as numbness and elevated sensory thresholds for temperature, pain, and other factors.

If it appears that these symptoms exist and may be related to vibration exposure, the employers should be advised that control may be insufficient. Because of possible association of back disorders with whole-body vibration, workers exposed to WBV should be advised during health surveillance about the importance of posture in seated jobs, and about correct lifting technique.

GENERAL POINTS ON VIBRATION

▨ A person who is warm and dry is less likely to suffer from HAV.

▨ HAV can cause damage to blood vessels and nerves in hands and fingers.

▨ The effects of HAV can be reduced if the work is done in short intervals.

▨ Anti-vibration gloves may not always protect against vibration.

RADIATION

RADIATION HAZARDS

Ionising radiation

Ionising radiation is produced by alpha and beta particles and gamma rays.

Harmful effects:

- Somatic – cell damage to the individual
- Genetic – cell damage to the children of the individual
- Acute – nausea, vomiting, skin burns and blistering, collapse and death
- Chronic – anaemia. leukaemia, other types of cancer.

Sources:

- Radon gas
- X-ray equipment
- Smoke detectors.

Non-ionising radiation

- Ultraviolet radiation – sun, arc welding
- Lasers – eye and skin burns, electricity etc.
- Infra-red radiation – fires, furnaces etc.
- Microwaves – cookers/ovens, mobile telephones.

Non-ionising radiation can harm the eyes and skin.

RADIATION PROTECTION STRATEGIES

Ionising radiation

Strategies involve:

- Risk assessment

- Shielding

- Time (reduced time exposure)

- Distance

- Training

- Personal protective equipment

- No food or drink consumption near exposed areas

- Signs and information

- Medical surveillance

- Maintenance and inspection controls

- Emergency procedures.

Non-ionising radiation

Strategies involve:

- Eye protection

- Skin protection (gloves and/or creams)

- Fixed shields and non-reflective surfaces

- Interlocking guards on microwaves.

STRESS

The reaction of the body to excessive mental pressure that may lead to ill-health.

CAUSES

- The job itself – unrealistic targets, boring, repetitive, insufficient training

- Individual responsibility – ill-defined roles, too much responsibility, too little control to influence outcome

- Working conditions – lack of privacy or security, unsafe practices, threat of violence, excessive noise

- Management attitudes – negative health and safety culture, poor communication, lack of support in a crisis

- Relationships with colleagues – bullying, harassment.

PREVENTION

- Take a positive attitude to stress

- Take employee concerns seriously

- Effective communication and consultation

- Develop a policy on stress

- Relevant training

- Employee appraisal system

- Discourage excessive hours at work

- Encourage lifestyle changes

- Monitor incidents of bullying etc.

- Avoid blame culture

- Set up a confidential counselling advice service.

Specimen Answers to NEBOSH Examination Questions

INTRODUCTION

This chapter is only relevant to those readers who are due to sit for NEBOSH examinations.

The accompanying textbook to this Revision Guide – *International Health and Safety at Work* – gives detailed advice on studying for NEBOSH examinations and some sample solutions to NEBOSH examination questions. It also provides a sample practical assessment. To gain the full benefit from this section of the Revision Guide, those chapters of the textbook should also be read. But full success in the examinations is only likely to be achieved if a good examination technique is adopted.

FEATURES OF A GOOD EXAMINATION TECHNIQUE

A. MNEMONICS

The use of mnemonics during your revision period can be very helpful in the examination. A mnemonic can be a sentence or a word that enables a person to remember a list of items. For example, the colours of the rainbow – red, orange, yellow, green, blue, indigo and violet – can be remembered by the sentence 'Richard Of York Gained Battles In Vain'. Similarly, the main factors in a manual handling assessment can be remembered by the word **TILE** – Task, Individual, Load and Environment.

The best mnemonics are those which students have devised for themselves. The reader is advised to invent his or her own set of mnemonics based on familiar words or sentences. Some examples of useful mnemonics are given below.

1. Management of international health and safety

Elements of a management system	Personal factors	Common topics in the management paper
Policy	Self-interest	Safety culture
Organisation	Hearing/memory loss	Training
Plan	Experience	Risk assessment
Monitor	Age	Information
Audit	Training level	Maintenance
Review	Health	Supervision
		Safe system of work

POPMAR, **SHEATH** and **STRIMSS** are stand-alone words.

(STRIMMS was devised by a student from the horticultural industry.)

2. Control of international workplace risks

Risk assessment stages (inc. COSHH and noise)	Machine hazards	Machine guarding
Assess the hazards	Entanglement	Fixed or fixed distance
Control the risks	Nips	Interlocked
Monitor the controls	Traps	Adjustable
Inform employees	Impact	Trip devices
Record and review the assessment	Contact or cutting	
	Ejection	

ACMIR can be remembered by 'All Colours Must Include Red'. **ENTICE** and **FIAT** are stand-alone words.

It is very important to stress that a mnemonic is only an aid to memory and not all the elements in a mnemonic may be relevant to a particular question. Where all or parts of a mnemonic are relevant to a question, you will need to expand on it in your answer.

For example, if the mnemonic, such as STRIMSS, reminds you that 'training and information' are relevant to the question, you should add a brief description of the training and information that you have in mind. Very few marks can be awarded for simply stating 'training and information'.

B. USE OF PERSONAL EXPERIENCE

The NEBOSH course covers much factual material which many students find difficult to remember. It is easier to remember this material if it can be related to personal experience and their own workplace. For example, most workplaces have some hazardous substances present, even if only as cleaning materials. These can be used to illustrate an answer involving hazardous substances (provided that the reference is relevant to the question!). Always try to use familiar examples of hazards and controls when answering examination questions involving those hazards and controls.

C. SIMPLE RULES WHEN TAKING THE EXAMINATION

1 Always arrive at least 10 minutes before the start of the examination.
2 Quickly read the whole paper before attempting any question.
3 Start each question on a separate sheet in the answer book and use both sides of the paper. The current NEBOSH answer book has the relevant question number printed on each page of the book.
4 Make brief notes at the beginning of the question page in the answer book.
5 Answer the easiest question first – this may not necessarily be the first one.
6 Read the question carefully and highlight the action verbs (e.g. outline, describe, state etc.). These action verbs are printed in **bold letters** in the question on the examination paper. It is common for candidates to answer the question that they would like to answer rather than the actual one on the examination paper.
7 Answer **all** parts of the question.
8 Attempt **all** questions – you can only get zero marks for an unanswered question.

9 Time is very limited during the examination and answers should be brief and to the point. Do not write all that you know about the topic being questioned or pad out your answer. You should spend no more than 25 minutes on the long answer question (covering about a page and a half) and 8 minutes on each short answer question (normally covering half a page). Most examination failures are caused by not allowing sufficient time to answer all the questions.

10 Finally, the answers should be easy for the examiner to read in terms of layout and standard of writing.

SPECIMEN ANSWERS TO NEBOSH EXAMINATION QUESTIONS

Some specimen answers are given below to past IGC1 and IGC2 examination questions. Detailed advice and an example of IGC3 – the Health and Safety Practical Application – are given in the textbook that accompanies this Revision Guide (*International Health and Safety at Work*).

It is recommended that, to gain full benefit from the specimen answers, the reader should attempt the questions initially without consulting the answers. The answers given are generally longer than would be expected in the examination because they have covered all possible options that are available in answering a particular question.

Finally, two long answer questions are provided for IGC1 and IGC2.

1. IGC1 – MANAGEMENT OF INTERNATIONAL HEALTH AND SAFETY

Question 1

(a) **Explain** reasons for maintaining and promoting good standards of health and safety in the workplace. **(8)**

(b) **Identify** sources of information that an organisation may use to help maintain and promote good standards of health and safety in the workplace. **(6)**

(c) **Outline** possible reasons why good standards of health and safety in the workplace may not be achieved. **(6)**

Answer:

(a) The reasons for maintaining and promoting good standards of health and safety in the workplace are identified as moral, social (and/or legal) and economic.

The moral argument is based on the need to provide a reasonable standard of care and to reduce the injuries, pain and suffering caused to workers by accidents and ill health, while the legal reasons centre on compliance with the law and ILO and other international standards to avoid criminal penalties and to comply with the employer's duty to take reasonable care of workers.

Social reasons are concerned with the need to provide a safe place of work, safe plant and equipment, safe systems of work, competent workers and a high standard of training and supervision.

The economic benefits would include a more highly motivated workforce, resulting in an improvement in the rate of production and product quality; the avoidance of costs associated with accident investigations; the avoidance of costs associated with accidents, such as the hiring or training of replacement staff and the possible repair of plant and equipment; securing more favourable terms for insurance; and maintaining the image and reputation of the organisation with its various stakeholders.

(b) The sources of this information that an organisation may use to help maintain and promote good standards of health and safety may be internal to the organisation and/or external to it.

Internal sources, which should be available within the organisation, include:

- accident and ill-health records and investigation reports;

- absentee records;

- inspection and audit reports undertaken by the organisation and by external organisations such as the national health and safety enforcement agency (HSE);

- maintenance, risk assessment (including COSHH) and training records;

- documents which provide information to workers;

- any equipment examination or test reports.

External sources, which are available outside the organisation, are numerous and include:

- health and safety legislation;
- UK HSE publications, such as Approved Codes of Practice, guidance documents, leaflets, journals, books and their website;
- European and British Standards;
- International Labour Organization (ILO);
- Occupational Safety and Health Administration (USA);
- European Agency for Safety and Health (EU);
- Worksafe (Western Australia);
- health and safety magazines and journals;
- information published by trade associations, employer organisations and trade unions;
- specialist technical and legal publications;
- information and data from manufacturers and suppliers; and
- the internet and encyclopaedias.

(c) Reasons why good standards of health and safety may not be achieved in the workplace include:

- a weak health and safety management structure together with either a lack or poor levels of health and safety competence;
- a lack of management commitment;
- the perception of a blame culture;
- poor morale among the workforce and a lack of motivation;
- poor selection procedures and management of contractors;
- frequent changes in the organisation and high staff turnover;
- a lack of resources, possibly due to a harsh economic climate;
- conflicting demands, with priority being given to production and financial targets and meeting deadlines;

■ poor communication and consultation with the workforce;

■ a failure to provide adequate training, leading to a lack of awareness amongst workers;

■ a failure to complete risk assessments and to produce safe systems of work and method statements; and

■ a lack of compliance with relevant health and safety law and the safety rules and procedures of the organisation.

Question 2

(a) **Identify** possible consequences to workers injured in an accident at work. **(4)**

(b) **Identify** possible costs to an organisation resulting from an accident at work. **(10)**

(c) **Outline** actions management may take to prevent similar accidents. **(6)**

Answer:

(a) The possible consequences to a worker injured in a workplace accident include pain and suffering and even disability or death with its resultant impact on family life. There could be a loss of earnings and future earning capacity following time off work and even loss of current employment. There may also be medical expenses resulting from the injury. There could also be a loss of confidence and motivation leading to social and psychological problems.

(b) Poor occupational health and safety performance results in additional costs to both public and private sectors of the economy of a country. Possible costs to an organisation resulting from an accident at work include both direct and indirect costs. Such costs include:

■ those associated with lost production and damage to products, such as production delays, the absence of employees such as a first aider who tends to the needs of the injured person, the recruitment and training of replacement staff and additional administration time incurred;

■ the need to pay the injured worker during their absence and to fund a temporary replacement, with the cost of additional training and possible extra overtime payments to other workers;

- repair of damaged buildings, plant, vehicles and equipment and the cost of the clean-up;

- accident investigation time and any subsequent remedial action required;

- an increase in insurance premiums due to claims on employers' and public liability insurance;

- fines and compensation awarded, and court and other legal representation costs; and any compensation not covered by the insurance policy due to an excess agreed between the employer and the insurance company;

- any attributable production and/or general business loss, including product or process liability claims; and

- intangible costs arising from a loss of business image and the detrimental effect on worker morale resulting in reduced productivity.

Some of these items, such as business loss, may be uninsurable or too prohibitively expensive to insure. Therefore, insurance policies can never cover all of the costs of an accident or disease, because either some items are not covered by the policy or the insurance excess is greater than the particular item cost.

(c) To prevent similar accidents, management could take actions such as carrying out a comprehensive investigation and communicating its findings to the workforce. This would indicate leadership and commitment to health and safety throughout and at all levels of the organisation. A review of the health and safety policy and existing related risk assessments and control measures would be essential. A programme of regular inspections and monitoring should be introduced and any defects found should be promptly rectified. There should also be a more effective standard of supervision and disciplinary action for non-conformance with set procedures. Effective consultation on a regular basis with the workers should be introduced, together with a programme of refresher training, not only on the operation of plant and equipment but also on general health and safety awareness.

Finally, there must be prompt investigation of all future incidents and accidents and reports made detailing any necessary remedial actions.

Question 3

(a) **Identify TWO** main purposes of first aid treatment. **(2)**

(b) **Outline** factors to be considered when carrying out an assessment of first aid requirements in a workplace. **(6)**

Answer:

(a) Two main purposes of first aid treatment are, firstly, the preservation of life and/or the minimisation of the consequences of injury until medical help is obtained and, secondly, the treatment of minor injuries that would not receive or do not need medical attention.

(b) The risk assessments should show whether there are any specific risks in the workplace requiring particular first aid provision. The following should be considered during an assessment of first aid requirements:

- the types of hazard and level of risk present, such as hazardous substances, dangerous tools and equipment, dangerous manual handling tasks, electrical shock risks and dangers from neighbours or animals;

- the distribution and composition of the workforce, including the special needs of workers such as trainees, young workers, nursing mothers and the disabled;

- the needs of travelling, remote or lone workers, such as the provision of personal first aid kits or mobile phones;

- the presence of members of the public on any site;

- the past history of accidents and their type, location and consequences;

- the number of trained first aid personnel and first aid facilities in relation to, for example, the size of the organisation;

- the proximity of the workplace to emergency medical services;

- the possibility of shared provision on multi-occupancy sites;

- the need to train the first aid personnel in special procedures; and

- the ability to provide continued cover over different shifts and for sickness, leave and other absence.

Question 4

(a) **Give** the meaning of the term health and safety 'audit'. **(2)**

(b) **Outline** the issues that need to be considered at the planning stage of an audit. **(4)**

(c) **Identify TWO** methods of gathering information during an audit. **(2)**

Answer:

(a) An audit is a systematic critical examination of a health and safety management system, involving a structured process for the collection of independent information with the aim of assessing the effectiveness and reliability of the total system and identifying its strengths and weaknesses in order to improve it where this is thought to be necessary.

(b) The issues that need to be considered at the planning stage of an audit begin with the selection of a competent audit team independent of the area to be audited. Then the objectives and scope of the audit must be agreed and audit questionnaires and checklists developed. At the same time, relevant guidance and standards that would be applied must be discussed and agreed. Decisions will need to be made about the level and detail of the audit before starting to gather information about the health and safety management of the organisation. Auditing involves sampling, so initially it will be necessary to decide how much sampling is needed for the assessment to be reliable. At the planning stage adequate resources and facilities, such as office space, must be allocated. Finally, timescales and feedback mechanisms must be agreed with relevant managers and employee representatives.

(c) The methods of gathering information during an audit include interviewing; reviewing and assessing written procedures; and observations of physical conditions and work activities to assess compliance with relevant health and safety standards and guidance.

Question 5

There has been a significant deterioration in the health and safety culture of an organisation.

(a) **Give** the meaning of the term 'health and safety culture'. **(2)**

(b) **Identify** the factors that could have contributed to the deterioration of the health and safety culture within the organisation. **(6)**

Answer:

(a) The health and safety culture of an organisation is the product of individual and group values, attitudes, perceptions, competencies and patterns of behaviour that determine the commitment to, and the style and proficiency of, the organisation's health and safety management.

(b) The factors that could have contributed to the deterioration of the health and safety culture within the organisation include:

- a lack of visible leadership and commitment at senior level;

- changes in the management structure or roles and in work patterns with a lack of effective communication prior to and during change;

- the fact that health and safety was not given the same priority as other objectives such as production or quality; lack of consultation with and involvement of the workforce;

- the absence of an effective health and safety management system;

- either a lack or poor levels of health and safety competence;

- poor selection procedures and management of contractors;

- a reduction in the workforce leading to work overload;

- a high staff turnover and external influences such as a downturn in the economy leading to job insecurity;

- the presence of a blame culture and/or peer pressure; and

- a deterioration in the standard of welfare facilities.

Question 6

Outline the main health and safety responsibilities of:

(a) employers; **(4)**
(b) workers. **(4)**

Answer:

(a) The main health and safety responsibilities of an employer are to:

- provide and maintain a safe workplace, including access and egress together with safe plant and equipment;

■ carry out risk assessments and introduce safe systems of work;

■ institute suitable occupational health and safety management arrangements appropriate to the working environment, the size of the undertaking and the nature of its activities;

■ provide, without any cost to the worker, adequate personal protective clothing and equipment which are reasonably necessary when workplace hazards cannot be otherwise prevented or controlled;

■ ensure the safe use, storage, handling and transport of articles and substances;

■ provide a safe working environment with adequate welfare facilities, including first aid;

■ prepare and when necessary revise a health and safety policy;

■ co-operate with and consult with workers;

■ secure competent health and safety advice;

■ co-operate with other employers at the workplace; and

■ provide information, instruction, training and supervision for workers.

(b) Workers have the responsibility to:

■ co-operate with their employer;

■ take reasonable care for their own safety and that of their fellow workers;

■ report accidents and any dangerous situations at the workplace;

■ not misuse any equipment provided for them;

■ follow site rules; and

■ not take alcohol or drugs during their working time.

2. IGC2 – CONTROL OF INTERNATIONAL WORKPLACE RISKS

Question 1

(a) **Outline** factors to consider when undertaking an assessment of health risks for a hazardous substance. **(8)**

(b) **Identify TWO** types of cellular defence mechanisms that the body has as a natural defence system. **(2)**

(c) **Give** the meaning of the term 'maximum allowable concentration'. **(2)**

(d) **Outline** factors that may reduce the effectiveness of a local exhaust ventilation system (LEV). **(8)**

Answer:

(a) The factors to consider when undertaking an assessment of health risks for a hazardous substance include:

- details of the process in which the substance is to be used;

- the hazardous nature of the substance, whether, for example, toxic, corrosive or carcinogenic, and its chemical, physical or biological hazardous properties;

- the form in which it is to appear in the workplace, for example as a dust or fume;

- the possible routes of entry of the substance, such as inhalation, absorption or injection;

- the possible ill-health effects of exposure to it;

- the frequency, duration and level of exposure and the number and type of persons who would be exposed;

- the existence of applicable standards such as work exposure limits and the suitability; and

- the adequacy of the control measures currently in place.

(b) Types of cellular defence mechanisms of the body include:

- the actions of white blood cells;

- the secretion of defensive substances;

- the prevention of excessive blood loss; and

- the repair of damaged tissue.

(c) The maximum allowable concentration is the concentration in air of a gas, vapour or substance which in general remains without harmful effects to both workers and their offspring even after repeated exposure and during a period of time up to an entire working life comprising 8 hours per day and 40 hours per week.

(d) The factors that may reduce the effectiveness of a local exhaust ventilation system (LEV) include:

- poor initial design;

- damage to the ducting; blocked, damaged, unsuitable or incorrectly installed filters;

- fan inefficiency, perhaps through wear or corrosion of the blades;

- inappropriate initial design, which may be exacerbated by process changes;

- unauthorised alterations or extensions, such as increasing the number of inlets;

- incorrect use, including a failure to position the hood close enough to the source of emission, a build-up of contaminant in the ducting and a blocked or obstructed outlet;

- scrubber saturation;

- incorrect settings, for example of the dampers; and

- a failure to introduce procedures for the regular maintenance, inspection and testing of the system.

Question 2

Construction workers can be exposed to hand–arm vibration during construction activities.

(a) **Outline** possible health effects of exposure to hand–arm vibration. **(8)**

(b) **Outline** control measures to reduce the risks to workers from exposure to hand–arm vibration. **(12)**

Answer:

(a) Hand–arm vibration syndrome (HAVS) describes a group of diseases caused by the exposure of the hand and arm to external vibration. Some of these are described as Work-Related Upper Limb Disorders (WRULDs), such as carpal tunnel syndrome and tenosynovitis. However, the best-known disease is vibration white finger (VWF), in which the circulation of the blood, particularly in the hands, is adversely affected by the vibration. The early symptoms are tingling and numbness felt in the fingers, usually some time after the end of the working shift. As exposure continues, the tips of the fingers go white and then the whole hand may become affected. This results in a loss of grip strength and manual dexterity. Attacks can be triggered by damp and/or cold conditions and, on warming, 'pins and needles' are experienced. If the condition is allowed to persist, more serious symptoms become apparent, including discoloration and enlargement of the fingers. In very advanced cases, gangrene can develop, leading to the amputation of the affected hand or finger.

(b) The following control measures should be taken to reduce the risks associated with HAV:

- avoid, whenever possible, the need for vibration equipment. Consider whether the process could be automated to avoid the use of vibrating equipment, or, if this is not possible, selecting equipment producing lower levels of vibration;

- undertake a risk assessment which includes a soundly based estimate of the employees' exposure to vibration;

- develop a good maintenance regime for tools and machinery. This may involve ensuring that tools are regularly sharpened, worn components are replaced or engines are regularly tuned and adjusted;

- introduce a work pattern that reduces the time exposure to vibration, perhaps by job rotation or the provision of frequent breaks;

- issue employees with gloves and warm clothing. There is a debate as to whether anti-vibration gloves are really effective but it is agreed that warm clothing helps with blood circulation, which reduces the risk of VWF. Care must be taken that the tool does not cool the hand of the operator;

- introduce a reporting system for employees to use so that concerns and any symptoms can be recorded and investigated; and

- ensure that workers are given information, instruction and training in the control measures to be adopted.

It is important that drill bits and tools are kept sharp and used intact – an angle grinder with a chipped cutting disc will lead to a large increase in vibration as well as being dangerous.

Question 3

Permission has been given for a mobile crane to be used on a construction site.

Identify checks that the driver should carry out before the lifting operation. **(8)**

Answer:

The checks that a mobile crane driver should carry out before the lifting operation include:

- a brief inspection of the crane and associated lifting tackle before each time it is used;

- the prevailing weather conditions;

- the availability of current inspection certificates for the crane;

- the condition of the ground on which the crane is to be sited to ensure it is firm and level;

- the availability and condition of the lifting accessories;

- the tyre pressures, where appropriate, are correct;

- outriggers have been deployed and are properly positioned;

- the load to be lifted is within the safe working load of the crane;

- adequate communication systems are in place and the driver and the banksman fully understand the signalling system to be used;

- the landing position is clear and there are no obstructions in the lifting path;

- no site personnel are working under the line of the lift;

- overhead obstructions or hazards must be identified; it may be necessary to protect the crane from overhead power lines by using goal posts and bunting to mark the safe headroom; and

- before lifting, that the hook is not attached to a fixed or anchored load.

Question 4

Identify welfare and work environment requirements that should be provided in a workplace. **(8)**

Answer:

The welfare and work environment requirements that should be provided in a workplace include:

- the provision of an adequate number of sanitary conveniences and washing facilities, including showers for both sexes. A good supply of warm water, soap and towels must be provided as close to the sanitary facilities as possible. The facilities should be well lit and ventilated and their walls and floors easy to clean. Hand dryers are permitted but there are concerns about their effectiveness in drying hands completely and thus removing all bacteria;

- a supply of drinking water must be readily accessible to the workforce and be adequate and wholesome;

- storage areas for clothing, including lockers and changing areas;

- facilities for the cleaning and replacement of working clothes;

- a rest room away from the working area with the facility for taking food and drink;

- the provision of effective and sufficient ventilation for the work area and an adequate heating system to maintain a reasonable temperature throughout the building;

- an adequate standard of lighting for the tasks being undertaken, with the emphasis on the availability of natural light;

- the condition of floors, stairways and traffic routes should be suitable for the purpose and well maintained;

- the provision of adequate space and suitable seating at the workstations; and

- the introduction of control measures to combat excessive noise.

NOTES

NOTES

NOTES

NOTES

NOTES

NOTES

NOTES